COLOUR, COLOUR MEASUREMENT AND COLOUR CHANGE

AF837718

Colour, Colour Measurement and Colour Change, the first new book in the expanded *Science for Conservators* series, explains the science of colour, colour measurement and colour change for conservators in a clear and comprehensible way, elucidating the topic for those with no scientific background.

The book explains how and why we see colours and how colour and colour change can be measured, as well as clarifying why these would be done in a conservation context. It then examines the ways in which colour can change – such as darkening, yellowing, fading, blanching and patination – illustrating these in different types of cultural heritage materials, including metals, varnishes, plastics, textiles and paints. The final chapter explores how colour change can be reduced in different types of storage and display settings, and, in particular, what can be done to protect against damage by light, damp and pollutants.

This book is an invaluable introduction to all aspects of the science of colour in conservation. It is suitable for students in undergraduate and postgraduate conservation programmes, as well as being a useful reference guide for practising conservators.

David Saunders, FSA FIIC, is Honorary Research Fellow and formerly Keeper of Conservation and Scientific Research at the British Museum. After postdoctoral research in chemistry, he began his conservation career in the Scientific Department at the National Gallery in London. He has twice been a guest scholar at the Getty Conservation Institute in Los Angeles and was also its inaugural Getty Rothschild Fellow. He has been a visiting professor at UCL and NYU. His research focuses on non-invasive analysis, the effect of the environment (particularly light) on museum objects and sustainable preventive conservation.

Science for Conservators
Series editor: Joyce H. Townsend

The *Science for Conservators* series is now being revised and fully updated, and it will be extended in scope. The original series has provided key basic texts for conservators throughout the world since its publication in the late 1980s. Scientific concepts are basic to the conservation of artefacts of every type, and these introductory volumes provide an essential theoretical background for conservators who have entered the field without scientific training. It is designed for pre-programme, student and personal study, and also serves as a checklist of scientific terms for those working in English as a second or third language.

An Introduction to Materials and Chemistry
Joyce H. Townsend

Colour, Colour Measurement and Colour Change
David Saunders

For more information about this series, please visit: www.routledge.com/Science-for-Conservators/book-series/SFC

Colour, Colour Measurement and Colour Change

David Saunders

Routledge
Taylor & Francis Group
LONDON AND NEW YORK

Designed cover image: Box of pigments and implements for painting on ceramics, produced by Hancock & Son and distributed by Reeves & Sons. [Source: Worcester / London, c.1880. Yale Center for British Art, Rare Books and Manuscripts, Watercolor Box 6. Paul Mellon Fund.]

First published 2025
by Routledge
4 Park Square, Milton Park, Abingdon, Oxon OX14 4RN

and by Routledge
605 Third Avenue, New York, NY 10158

Routledge is an imprint of the Taylor & Francis Group, an informa business

© 2025 David Saunders

The right of David Saunders to be identified as author of this work has been asserted in accordance with sections 77 and 78 of the Copyright, Designs and Patents Act 1988.

All rights reserved. No part of this book may be reprinted or reproduced or utilised in any form or by any electronic, mechanical, or other means, now known or hereafter invented, including photocopying and recording, or in any information storage or retrieval system, without permission in writing from the publishers.

Trademark notice: Product or corporate names may be trademarks or registered trademarks, and are used only for identification and explanation without intent to infringe.

British Library Cataloguing-in-Publication Data
A catalogue record for this book is available from the British Library

ISBN: 978-1-032-50173-4 (hbk)
ISBN: 978-1-032-50171-0 (pbk)
ISBN: 978-1-003-39718-2 (ebk)

DOI: 10.4324/9781003397182

Typeset in Times New Roman
by Newgen Publishing UK

Contents

Illustrations

Figures

Tables

Foreword

The series was first published in 1982 as a series of three books aimed at conservation students and conservators who had come into the field without any scientific training beyond their high school years. It was intended to develop into some six or seven topics covering a broader spectrum of knowledge than the first three, which concentrated greatly on chemistry. The series of three was reprinted in 1987 and produced by Routledge as a second edition in 1992. It has been in steady demand ever since, despite the text being unchanged since the early 1980s.

The collective memory of all those involved with the first edition is hazy except on one point: the original series was intended to have a book about colour, dealing with colour measurement, which was then quite unusual in the field of conservation (because the calculations were laborious and done by hand) and perhaps colour change. Or perhaps there should have been a book about deterioration and colour change: recollections vary. (And recollections ran out before the seventh proposed title: the sixth should have been something to do with examination and analysis. In the 2020s, that seems far too large a topic for one book, and there are excellent books to be found on so many aspects of this subject, many of them written by conservation professionals for their fellows.) No other book on colour, aimed specifically at the conservation profession and acknowledging that most of its members are not graduates in physics or mathematics, has been published since. Conservators have had to grapple with rather dense books on colour science, aimed at industrial or academic scientists – until now.

This book may still be one of the longer books in the new series, because it introduces a great many concepts concerning colour and appearance, but it is rooted in materiality, and the conservation and preservation of objects whose appearance matters. Nearly every object encountered by a conservator provokes questions such the following. What am I seeing? How has the appearance of this object changed over time? How much will my planned treatment change its appearance? Can my treatment take it closer to its colour when new? If I choose non-intervention, will its colour change in future? This book is intended to equip conservation professionals to answer these core questions in practice.

It has been structured so that it is possible to concentrate on the early chapters and then on *one* of the later ones, such as colour measurement, because that is the focus of the next project, treatment or report.

The fully revised and updated *Science for Conservators* series is intended to serve as a basic primer, and an aid to the assimilation of concepts that will become embedded in the thought processes involved in decision-making by conservation professionals. The series is targeted at several groups: those about to enter a conservation programme without personally having a scientific background; conservation professionals in training and with a similar background; those who are establishing themselves in their challenging and rewarding field; professional conservators who may want to check that they are using a scientific term to mean the same thing a scientist would assume it to mean, rather than its meaning(s) in everyday conversation; and those who have studied the natural sciences or even the conservation of artefacts in another language than English, but are now working, writing reports, and publishing in English.

When scientific terms appear or are defined for the first time, they are **emphasised thus** for easy reference. A full index of such terms is included at the end of each book.

Joyce H. Townsend
Series editor
August 2024

Acknowledgements

I would like to thank series editor and long-time colleague Joyce Townsend for offering me the opportunity to write this volume on colour – a subject I have long thought lacked a basic text aimed at conservators – for inclusion in the *Science for Conservators* series.

Many people have contributed to this book, sometimes inadvertently: the encouraging and thoughtful comments of the reviewers, the discussions over coffee with colleagues at the National Gallery, London, the British Museum and the Getty Conservation Institute, the long-term partnership with Jo Kirby in all matters relating to colour change and the detailed reading of all or part of the text by Joyce Townsend and Catherine Higgitt. The team at Taylor & Francis (Marcia Adams and Alison Macfarlane) have expertly and patiently steered me through the publication process and I am grateful to Susan Dunsmore for her razor-sharp yet sensitive copyediting of the text. Finally, my family and friends have cheerfully borne my physical and mental absences from 'real life' while researching and writing this book.

Introduction

Most conservators will have undertaken a colour vision test at some stage during their career, perhaps as a student or when applying for a job. The ability to see and distinguish colours is of great importance in our profession, even for those who work with objects that are – at least at first sight – less colourful. While the colour of some objects may result simply from the intrinsic properties of its component materials, colour can also play many different roles: it can, for instance, be decorative, symbolic, emotional or representational.

An understanding of some of the scientific principles of colour helps conservators to appreciate: the nature of objects; the way in which their colour may have changed (or may change in the future); the potential effects of treatment; and the steps that can be taken to promote the preservation of coloured materials.

This series of books on *Science for Conservators* offers the opportunity to draw together in one volume many of the colour-related scientific principles and techniques that are relevant to conservators. The choice of what to include has necessarily been subjective, but the seven chapters in this book present:

- an understanding of how our eyes and brains see colours;
- the chemical and physical principles that cause materials to appear coloured;
- how we classify and measure colour and differences between colours;
- the mechanisms by which colours are lost or change;
- what types of change in colour we might expect to see in objects;
- the reasons why we measure colour and colour differences;
- preventive measures we might take to mitigate such changes in colour.

Currently, information relevant to these topics is either found in more general physics and chemistry texts or is covered topic-by-topic in articles in conservation and other journals. While the aim of this volume is to gather pertinent information in one publication, most of the texts and articles from which material is drawn are referenced, allowing the reader to explore specific topics in more depth than is possible in this introduction to colour. Where possible, recent publications have been cited, which – hopefully – give a view of previous studies in a particular area.

DOI: 10.4324/9781003397182-1

Since each chapter may be read in isolation, there is a degree of repetition from one to another, minimised by cross-referencing to the section with the fullest discussion.

Those looking for further general background reading might consult some of the texts most frequently cited in this volume:

- *Color Science and the Visual Arts* (Berns 2016);
- *Billmeyer and Saltzman's Principles of Color Technology* (Berns, Billmeyer and Saltzman 2000);
- *Colour Chemistry* (Christie 2015);
- *Measuring Colour* (Hunt and Pointer 2011);
- *Color Science in the Examination of Museum Objects* (Johnston-Feller 2001);
- *The Physics and Chemistry of Color* (Nassau 2001).

1 The perception of colour

How we see colours

Colour surrounds us to the extent that we take it completely for granted, but why does one object appear red and another green? The answer lies partly within the objects, which we will investigate in Chapter 2, and partly within us, in the way our eyes and brain gather information from the object. As we will see, the perception of colour relies on both the light that reaches and stimulates our eyes and the way in which these stimuli are processed and interpreted by our brain.

1.1 Electromagnetic radiation

We need to begin by looking at light in a little more detail. We use the term 'light' to refer to radiation in a particular portion of the so-called **electromagnetic spectrum**, distinguished by the fact that it is the region to which our eyes are sensitive. We may well be familiar with other types of radiation from this spectrum – for example, X-rays, ultraviolet or infrared radiation – all of which are shown in Figure 1.1.

We can consider radiation as a wave and the scales along the top and bottom of Figure 1.1 refer to the **wavelength** of each type of radiation, that is the distance between successive peaks of the waveform (see *Wavelength, frequency and energy* box). The wavelength (denoted by the Greek letter λ, pronounced 'lambda') is measured in metres, or fractions of a metre for convenience; thus, visible light has a range of wavelengths from approximately 400–700 nanometres (nm).[1] When considering the properties of visible light and its perception or measurement, it is common to see graphs in which the horizontal axis plots this 'spectrum' of wavelengths from around 400–700 nm, see, for example, in Figures 1.3 and 1.4 later.

1 A nanometre is 1/1,000,000,000 of a metre – sometimes also written as 1×10^{-9} m.

DOI: 10.4324/9781003397182-2

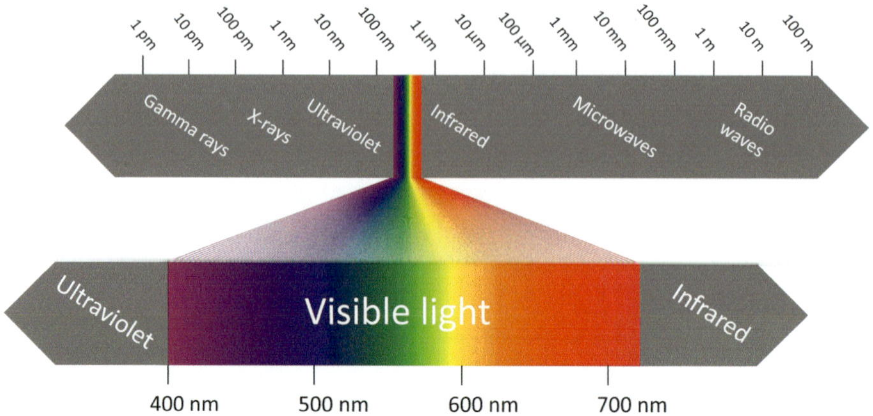

Figure 1.1 The electromagnetic spectrum showing the visible region, which is responsible for colour, in more detail

Wavelength, frequency and energy

Different types of electromagnetic radiation are also defined by their frequency (denoted by the Greek letter ν, pronounced 'nu'). As, in a vacuum, all electromagnetic radiation moves at the so-called speed of light (c), the number of wave peaks that pass a given point per second will increase as the wavelength becomes shorter. This can be expressed mathematically as:

$$\nu = c/\lambda$$

In addition, the energy (E) associated with different types of radiation is connected to their frequency/wavelength through Planck's constant (h):

$$E = h\nu \quad \text{or} \quad E = hc/\lambda$$

This implies that as wavelength decreases, energy increases. So, in Figure 1.1, cosmic rays will have higher energy than X-rays and ultraviolet radiation higher energy than visible radiation.

In Figure 1.1 the different wavelengths of visible light are depicted as colours, but it must be remembered that colour is an attribute we associate with the sensation produced by light of a certain wavelength and not an intrinsic property of the radiation itself. That said, these colour descriptors are a useful shorthand when describing radiation of different wavelengths and are much used in the sections that follow.

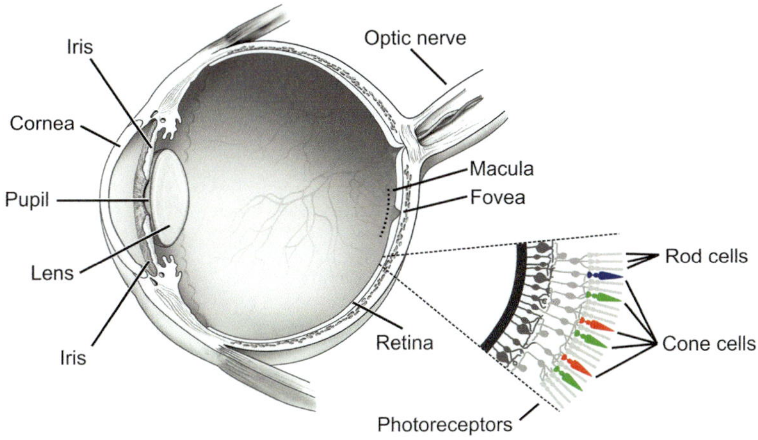

Figure 1.2 Cross-section of the human eye with a detail showing the photoreceptors in the retina

1.2 Colour vision: the role of the eye

While the light reaching the eye, and specifically the range of different visible wavelengths, bears information about colour, it is the way in which this is sensed by the eye that is the first step in our perception of colour. The rear surface of our eyes, the retina, onto which an image of a scene is focused (upside-down, in fact) by the lens, contains a number of different types of light-sensitive cells (Figure 1.2). Among these, the so-called cone cells are critical for **colour vision** (the rod cells shown in Figure 1.2 are principally responsible for night vision while other non-visual cells are involved in the body's circadian – day/night – rhythms). In the majority of human eyes there are three types of cone cell, which are – broadly – sensitive to the red, green and blue regions of the visible spectrum, sometimes also referred to as the long, medium and short wavelength cones, respectively. All three types of cone cell contain a modified version of the light-sensitive chemical rhodopsin, but the modification differs in the red, green and blue cones, making them responsive to these different wavelengths of light. Figure 1.3 shows the photosensitivity of the three receptors to different wavelengths in an *average* viewer (there are some differences between individuals).

If blue light with a wavelength of 400–450 nm enters the eye, then it will be the blue cone that is most stimulated, while red light with a wavelength of *c*.650 nm would elicit the greatest response from the red cone. For light between these extremes, the colour perceived will depend on the ratio of stimulation of the three cones in response to the range of wavelengths present.

Considering the effect of blue, red or other coloured lights on the cone cells is a useful tool to explain the response of the eye, but in most circumstances we will view objects under what we term 'white' light sources and the colour(s) we perceive will derive from the interaction of the object with that white light.

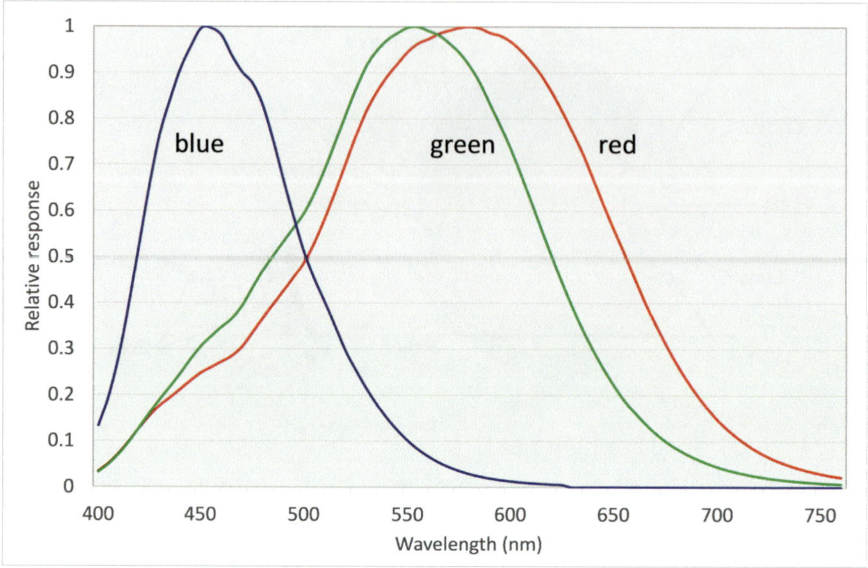

Figure 1.3 Average spectral sensitivity curves for the three cone sensors in the human eye

1.3 White light

The classic physics experiment of passing white light through a prism shows us that white light is composed of component colours across the visible spectrum. But experience also tells us that not all 'white' lights are identical – we would consider both domestic (or museum) electric lights and daylight to be white but sense that their colours often differ greatly. Graphically, this can be represented by showing the contributions of the various component wavelengths to the overall spectrum of a white light source. Two examples are illustrated in Figure 1.4: (a) is a spectrum of the now almost obsolete tungsten incandescent lamp; and (b) is a spectrum representative of daylight. While definitions of white light often refer to the presence of all the wavelengths in the visible spectrum at equal intensity, this is clearly not the case in either of the spectra in Figure 1.4. Although less precise, a useful definition of white light is that it stimulates all three of the cones to a similar – if not equal – extent. An explanation of why our visual system regards both incandescent light and daylight to be white despite their obvious differences is given in Section 1.6.3.

Although not relevant in the context of colour, we can also see that the two spectra in Figure 1.4 demonstrate rather different emissions in the ultraviolet region (the 300–400 nm region which is included here), which will be of significance when we come to consider the damaging effects of different white light sources, see Chapters 5 and 7.

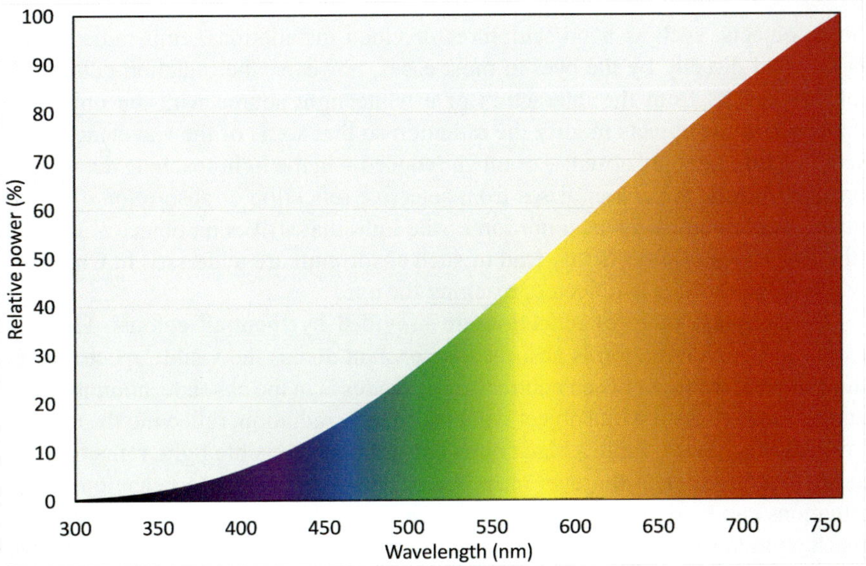

(a)

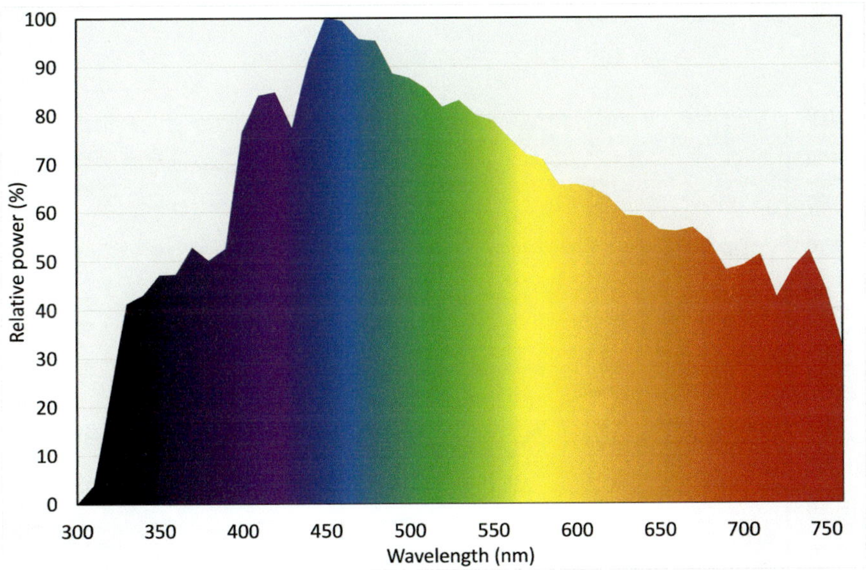

(b)

Figure 1.4 Spectral power distributions for: (a) an incandescent lamp; and (b) north sky daylight in the northern hemisphere

1.4 The colour of objects

Some objects, such as neon sculptures or video installations, emit radiation that is detected directly by the eye. In most cases, however, the radiation detected by the eye results from the interaction of a white light source with the object; the materials in the objects modify the radiation so that some of the wavelengths present in white light are absent – or much reduced – in the light reaching the eye. For opaque objects, the colour arises from selective reflection or absorption of certain visible wavelengths. Thus, a portion of the light that strikes an object is absorbed (the properties of objects that lead to such absorption are addressed in Chapter 2) while the remainder is reflected, reaching the eye.

The simplest cases of reflection are provided by 'neutral' colours. Here, the extent to which radiation is absorbed is constant across the visible spectrum, and the main difference between neutral (grey) shades is in the absolute amount of radiation reflected. So, a white object will absorb little radiation, reflecting the majority towards the viewer, while a black object absorbs most visible light, reflecting only a small proportion to the eye; grey objects show intermediate behaviour. These situations can be represented graphically in the 'reflectance spectra' for white and black pigments (and the mixed grey) shown in Figure 1.5. Theoretically, a 'perfect' white or black would reflect 100 per cent or 0 per cent respectively of the light striking the material, but here the reflectance approaches but does not reach these limits. More significantly, for each neutral colour all wavelengths are reflected

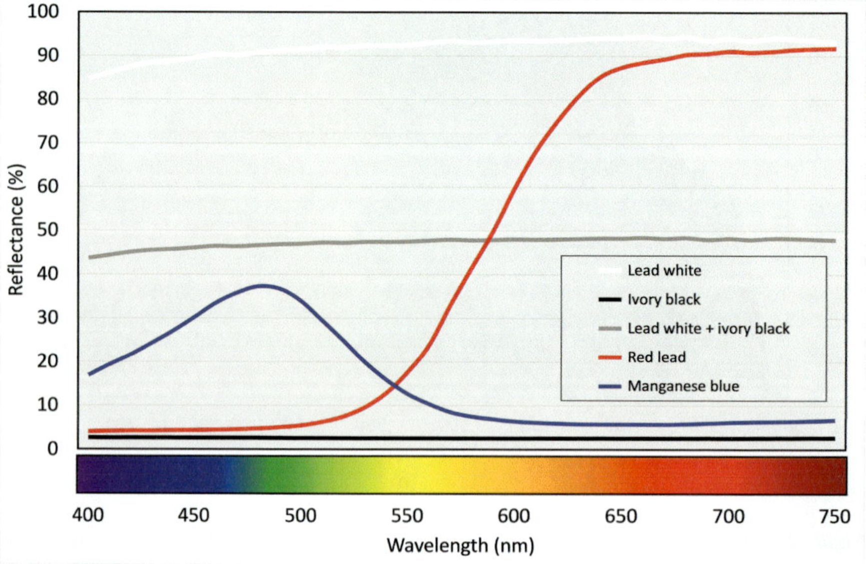

Figure 1.5 Reflectance spectra for neutral colours (white, grey and black) and red and blue colorants

nearly equally, so there is no, or very little, change in the balance of colours in the light reaching the viewer.

Most materials, however, interact with the illuminating light in a manner that alters the balance between wavelengths. Two examples are given in Figure 1.5, which shows the reflectance spectra for two pigments: red lead and manganese blue. When illuminated with white light, the red pigment absorbs radiation in the blue and green regions of the spectrum, reflecting the orange and red light to which it owes its colour. Conversely, manganese blue absorbs red and orange radiation, reflecting blue and some green light.

In transparent or translucent materials colour is again the result of selective absorption. For example, when we observe a piece of green glass against a white light, most of the radiation (including, blue, yellow, orange and red light) is absorbed, and only green light is transmitted to the viewer, giving the glass its characteristic colour. The **transmittance** of the material can be plotted against the wavelength in the same way that **reflectance** is plotted against the wavelength in Figure 1.5.

1.5 Gloss and scattering

We have seen that through absorption, reflection or transmission the light striking an object can interact to give rise to colours characteristic of its materials. There are, however, other ways in which light interacts with objects. Two phenomena that are important in determining colour occur at the boundaries between materials.

The first, **gloss**, depends on the properties of the outer surface of the material – generally the interface between the object and the surrounding air. While much radiation will pass into the object and interact through absorption, reflection or transmission, some light is reflected at the surface (see the *Reflection and refractive index* box). If the object is extremely smooth or polished, then the surface acts like a mirror, reflecting the light in a single direction, which is related to the angle from which the surface is illuminated, see Figure 1.6.

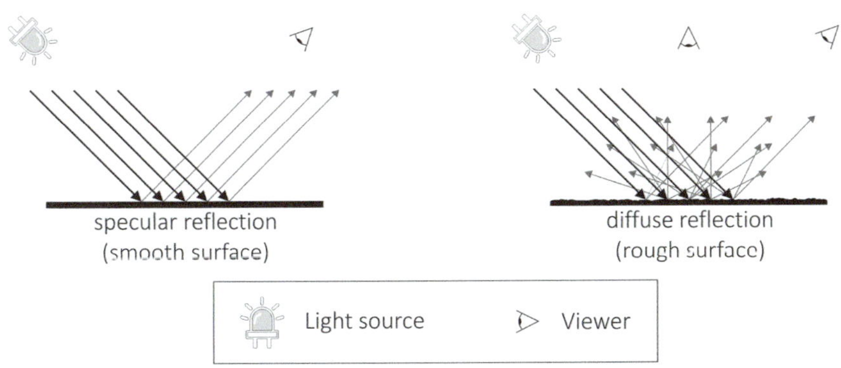

Figure 1.6 Schematic representation of specular reflection (gloss) and diffuse reflection

Reflection and refractive index

The laws of reflection state that the angle at which radiation is reflected from a surface is identical to the angle at which it is incident on the surface: that is, $\theta_I = \theta_R$ in the diagram below. (The Greek letter θ, pronounced theta, is used to describe the angle, as shown here.)

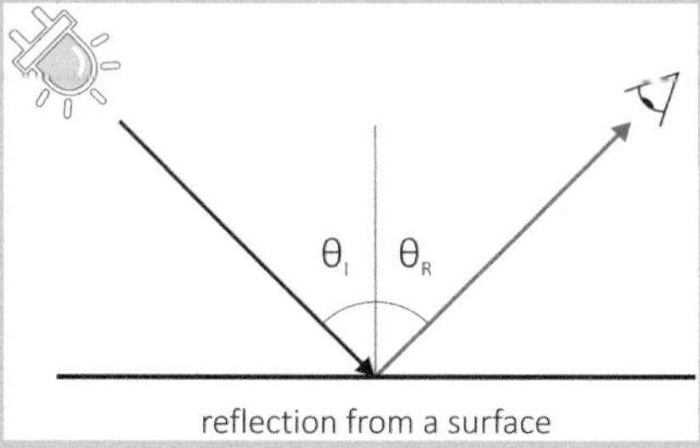

The refractive index of a substance (n) is related to the speed at which light travels in that substance compared to the speed of light in a vacuum. As light passes from one substance to another its course is changed (it is refracted). The change in angle (shown below) is determined by the difference in the refractive indices, where:

$$n_1 \sin\theta_1 = n_2 \sin\theta_2$$

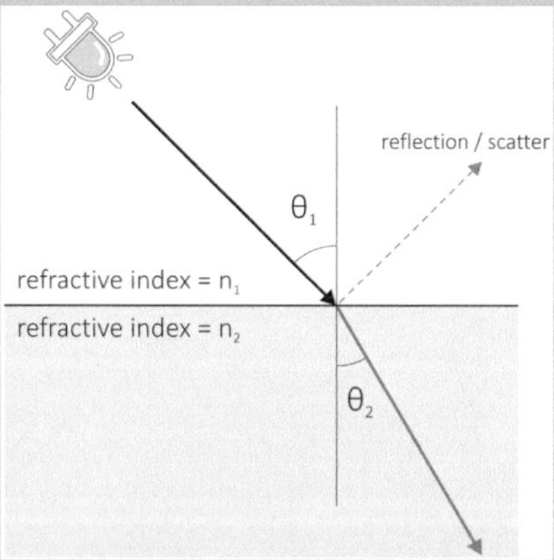

The refractive index of a material may vary with wavelength, which is why white light is split into its component colours when passing from air into a glass prism.

This type of surface reflection is termed 'specular reflection' or gloss. Because specular reflection occurs at the surface, before the light has interacted with the material, its colour is that of the illuminating light rather than the object and an image of the light source can sometimes be seen on its surface.

At rougher surfaces the incident light is reflected in myriad directions, depending on its orientation at each point, see Figure 1.6. This behaviour is termed '**diffuse reflection**'; no image of the light source is formed, but the reflection of white light in every direction can serve to desaturate the appearance of the underlying colour (Figure 1.7).

The second phenomenon is **scattering**, which can occur within the object, for example, at the boundary between pigment particles and a binding medium. The degree to which light is scattered is related to the refractive indices of the two materials, see the *Reflection and refractive index* box. If the indices are comparable, there is little scattering and the particles appear translucent or transparent (Figure 1.8). In a layered system, this allows the material beneath to be seen. In contrast, if the indices differ greatly, scattering at the boundaries makes the particles

Figure 1.7 Matt (top) and gloss (bottom) paint films. Note the desaturation of the colour by reflected white light in the gloss sample

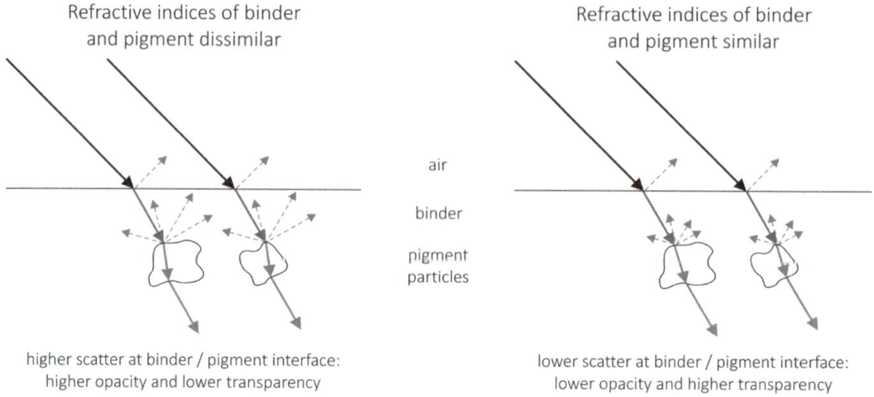

Figure 1.8 Schematic representation of the effect of refractive indices on the passage of light through a binding medium and pigment particles

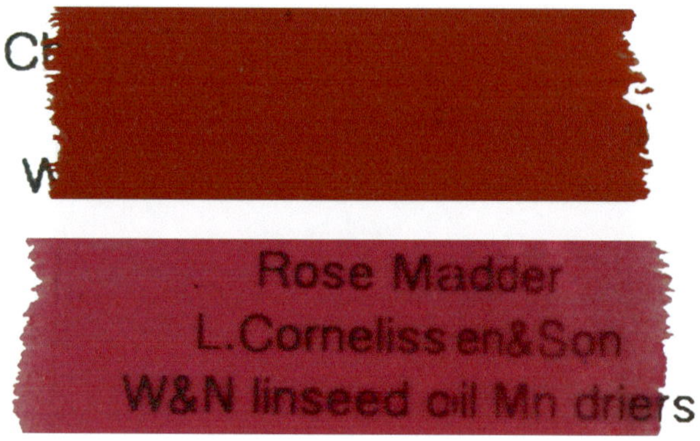

Figure 1.9 Opaque (top) and transparent (bottom) paint films showing their different hiding powers

visible and the layer has greater opacity (sometimes referred to as **hiding power**) (Figure 1.9).

Hiding power is often assessed by applying the layer under study to an inert card that has both black and white segments and can be quantified using standard procedures, such as those described in the American Society for Testing and Materials (ASTM) standard test procedure ASTM D6441 (ASTM 2016).

A further potential interaction between radiation and materials that can give rise to colour is the phenomenon of **luminescence**, where radiation at one wavelength is absorbed and re-emitted at a longer wavelength (with lower energy). This will be considered in more detail in Chapter 2.

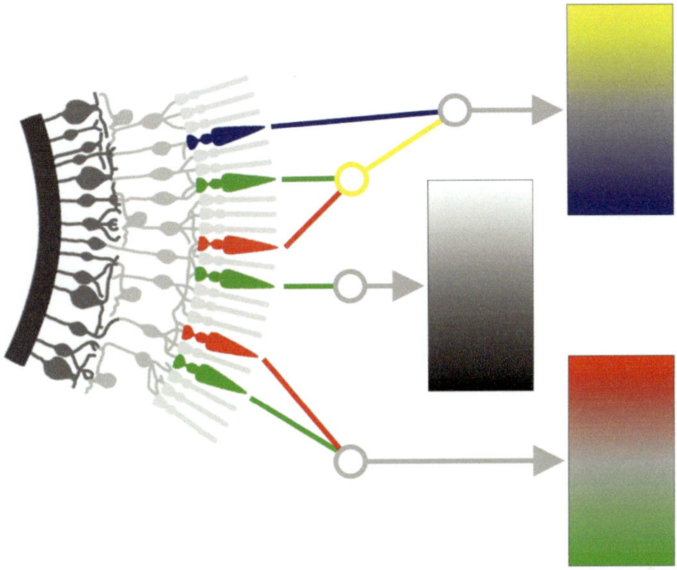

Figure 1.10 Schematic representation of the processing of the signals from the red, green and blue cone sensors to generate three opponent signals: dark–light, red–green and yellow–blue

1.6 Colour vision: the role of the brain

Having seen how interactions between white light and objects determine the range of wavelengths of radiation reaching the eye, we can return to the way in which we perceive colour, considering how the raw signals produced by the red, green and blue photoreceptors in the eye are further processed and analysed. Some earlier theories of colour vision – for example, the trichromatic theory propounded by Young and Helmholtz – assumed that the cone signals were used directly in the perception of colour. However, the currently accepted model (which is also the basis for modern methods of categorising and measuring colour) is founded on the **opponent colour principle**. This model, proposed by Hering in the late nineteenth century, establishes three opponent scales: black *vs* white; green *vs* red and blue *vs* yellow.

Research suggests that the signals from the red, green and blue cones are processed within the nerves cells (neurons) in the eye before these three opponent signals are transmitted to the brain. A simplified description of this processing is given here and illustrated in Figure 1.10; for more details, see Hunt and Pointer (2011: 8–9, Figure 1.4).

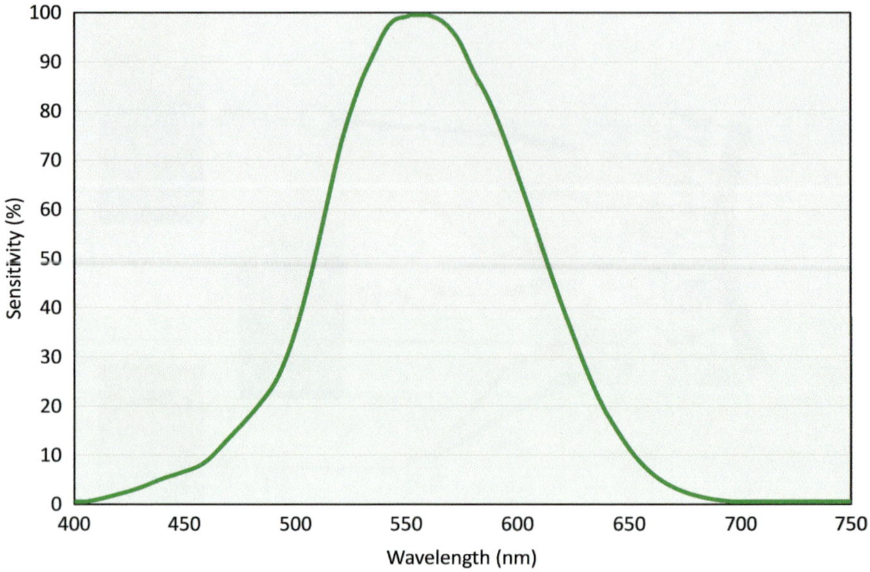

Figure 1.11 The luminous efficiency function, the v(λ) curve

The dark *vs* light signal interprets the brightness[2] of the colour, but is essentially achromatic – that is, it bears no information on hue or saturation but simply places it on a neutral scale from black to white. The signal derives predominantly from the green cones (with some contribution from the red cones and the rod cells responsible for night vision) and thus conveys darkness to lightness on an opponent scale. This close correlation between the stimulation of the green cone and our perception of lightness relates to the position of green radiation at the centre of the range of wavelengths to which our eyes are sensitive. The sensitivity of the eye to various visible wavelengths under normal lighting conditions – centred on the green region – is shown in Figure 1.11 and is termed the luminous efficiency function, sometimes written as v(λ).

The remaining colour information is encoded in the other two opponent signals. The green *vs* red signal is generated in the neurons of the eye by contrasting the responses of the green and red cones, see Figure 1.10. The final opponent channel (yellow *vs* blue) derives from a comparison of the response of the blue cone with the combined responses of the green and red cones (as the yellow portion of the spectrum coincides with wavelengths at which both these cones are stimulated).

2 The terms brightness and lightness are often confused. Either can be used correctly to express the amount of light reaching the eye from a scene, but only lightness is appropriate to describe the increased reflectivity of a paler object.

These opponent colour signals are combined with the brightness signal by the brain to form an overall image of a scene.

This is not, however, the end of the process and it would be a mistake to consider the combination of eye and brain as akin to an instrumental colour measurement system. Our visual system can adapt to changes in viewing conditions using direct information from the scene and can interpret the information conveyed from the eye using both contextual and experiential knowledge. The result is that our visual system is extremely powerful and sophisticated, but much more subjective than the instruments we might use to measure colour.

1.6.1 Metamerism

The phenomenon that demonstrates the fallibility of our visual system is **metamerism**. Sometimes, when two colours with different reflectance properties are viewed under a particular light source, they may give rise to identical visual signals. But when seen under a second light source, with a different distribution of energy across the spectrum, the signals from the two colours may differ, that is they appear to match under one light source but not under another. In conservation, metameric mismatches of this type are most common for blue pigments – frequently arising from their varying reflectance properties at the red end of the spectrum. This has led to difficulties when, for example, good retouching matches have been made in a daylit studio that do not match when seen under electric light in display or study spaces (Staniforth 1985). Film or digital images can also show metameric mismatches, particularly blue retouches on painted surfaces. This phenomenon is, however, due to the difference between the colour response of the film or camera and that of the eye, rather than to differences in the spectra of light sources.

1.6.2 Dark adaptation

Returning to the ways in which our visual system can adapt to external circumstances, we have probably all experienced our ability to adapt to falling light levels while still maintaining colour vision, until, that is, a point is reached where we can no longer distinguish colours and see only by **achromatic night vision**, also known as **scotopic vision**. Physical and biochemical processes contribute to this adaptation. An involuntary dilation of the pupil allows more light to enter the eye, an increase in the concentration of rhodopsin offers greater sensitivity, and the transmission of signals by the retinal neurons is accelerated. These changes commence quickly but, depending on the new level, full **dark adaptation** can take up to 10–15 minutes (Boyce 2014: 59, Figure 2.12).

If the light level is too low, the cone responses drop to a point where the visual system relies on the more sensitive rod cells and vision becomes achromatic, that is, colour-free. There is, however, no sharp transition from full colour (photopic) vision to achromatic (scotopic) vision, but our ability to distinguish colours and the relationships between them gradually decreases. Three phenomena of note in this context are an apparent change in the relative lightness of different colours (the Purkinje effect), a shift in the perceived hue of colours (the Bezold-Brücke effect: Pridmore 1999), and a general reduction in colourfulness of scenes (the Hunt effect: Hunt and Pointer 2011: 303).

1.6.3 Colour adaptation

Colour adaptation or **colour constancy** is the visual system's response to the issue we saw earlier, that of the variable nature of white light. If we look at a piece of white paper – for instance, a printed page – indoors or outdoors, we still interpret its colour as white, despite the very different distributions of wavelengths likely to be present (see the examples in Figure 1.4). The simple electrochemical responses of the three cones will continue to be dictated by the spectrum of the light, but the processing of the resulting signals by the brain can compensate for the effects of the illuminant.

The principal difference between the spectra of the two light sources in Figure 1.4 is that the incandescent lamp emits much less radiation in the blue region and much more in the red when compared to daylight. This difference is quantified by defining a **colour temperature** for different types of white light (see *Colour temperature* box).

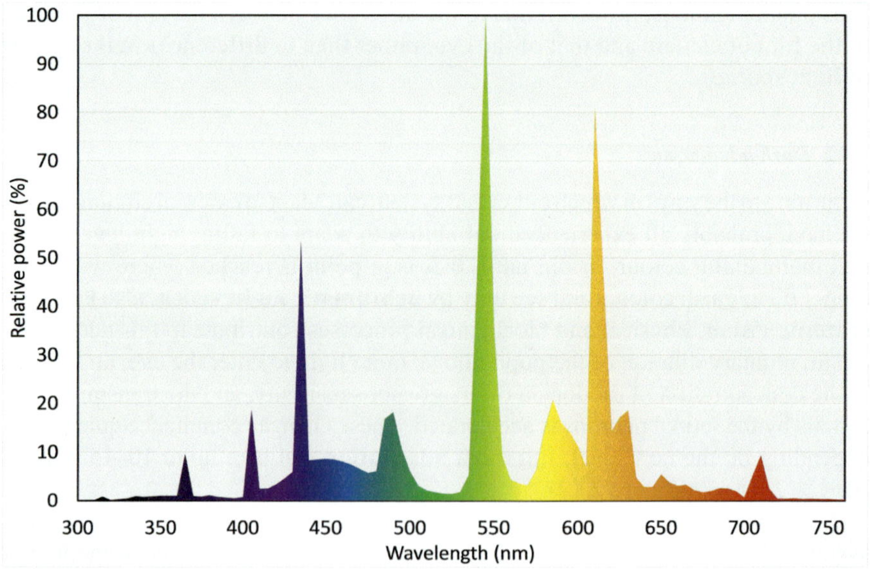

Figure 1.12 Spectral power distribution for a high efficacy, white fluorescent lamp

Colour temperature

White light sources are frequently classified by their colour temperature, defined as the temperature to which it would be necessary to heat a perfect emitter of radiation (often called a black body or Planckian radiator) for it to produce light whose spectrum matches that from the light source we are considering.

As metals (which are good approximations of black body radiators) are heated, they first emit red, then yellow light, while at yet higher temperatures they produce a bluish light. Light sources such as incandescent or halogen lamps, which create light by heating a metal filament, behave like black body radiators (see Figure 1.4 (a)). For fluorescent (see Figure 1.12) and LED lights, which produce white light by emitting in discrete regions of the spectrum, the correlated colour temperature (CCT) – the term used when sources are not similar to black body radiators – is the colour temperature of the black body with the closest colour match to the source under consideration.

The CCT is measured in Kelvin (abbreviated as K), the temperature scale used in scientific contexts: one Kelvin equals one degree Centigrade (1°C), but the zero point on the Kelvin scale (termed absolute zero) equates to *c*.-273°C.

Light sources with lower colour temperatures contain a higher proportion of red radiation, while those at higher colour temperatures contain more blue light. The approximate equivalence between colour temperature, colour and some common light sources is show in the scale below.

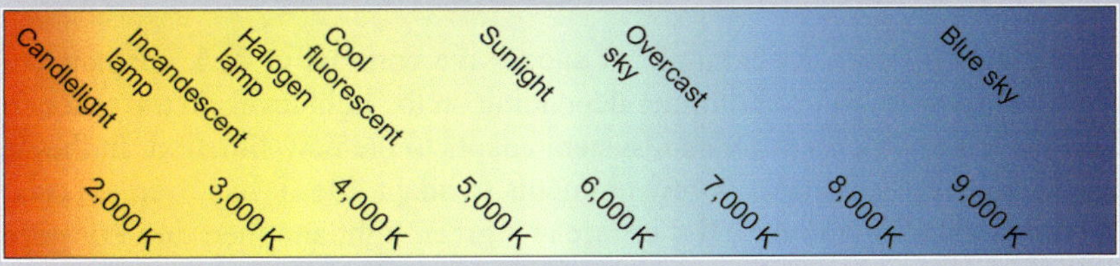

A very simplistic view of colour constancy is that it serves to balance the red and blue signals so that areas of a scene that the brain interprets as being white or neutral coloured appear so, whatever the light source. The 'white balance' function present on many cameras attempts to emulate this process, either automatically or by prompting us to image a white or grey area.

Advantage can be taken of the fact that stimulation of the three cone sensors produces the sensation of white light. Figure 1.12 shows the spectrum of a fluorescent lamp that emits only in very narrow regions of the visible spectrum. However, these emissions have been chosen deliberately to correspond to the sensitivity of human cone cells, so that the light appears white. While such lamps are quite energy-efficient (although less so than modern LEDs), the appearance of colours can be distorted under their light.

Colour adaptation will not, however, compensate for strongly coloured light sources and we will likely experience difficulty interpreting colours under these circumstances – for example, under yellow (low pressure) sodium lamps that are still occasionally found in streetlights, or under the red 'safelight' in a photographic darkroom, although both are increasingly anachronistic examples.

1.6.4 Simultaneous contrast

Our perception is also influenced by the lightness and colour of the area surrounding the subject at the centre of our field of view. In the upper part of Figure 1.13 the two grey discs are identical, but that on the left appears darker because of its paler surroundings, while in the lower part of the figure the coloured discs are identical, but do not appear so as a result of their different surroundings. Since we infrequently see colours in isolation, we need to be aware of this influence of the surroundings on perceived colour (especially when inpainting a loss on an object).

Simultaneous contrast can also be used to advantage, for example, to emphasise adjacent colours, as proposed by the French colourist Michel Eugène Chevreul, whose complementary colour system we will encounter in Chapter 3, and which was widely employed by impressionist and post-impressionist artists (Bomford et al. 1990: 76–83).

1.6.5 Successive contrast

We have already seen an example of successive contrast in dark adaptation – a room can appear brighter or dimmer depending on the light levels in the space from which we have come as our visual system adapts to the new situation. In the same way, the visual system takes a few moments to adapt after it has been exposed to a strongly coloured stimulus. If we stare at a green light and then turn our gaze to a neutral white or grey surface, for a short time it will appear magenta in tone, the colour that is complementary to green (see Section 1.8 for a definition of complementary colours).

1.7 Colour appearance models

While dark adaptation and colour constancy are useful for us as viewers, they – and the phenomena of simultaneous and successive contrast – highlight the subjective nature of our visual system, pointing to the need for methods other than human vision to assess and quantify colours objectively; we shall consider such methods of objective colour measurement in Chapter 3.

There have, however, been attempts to develop a mathematical model that takes these and other visual parameters that affect colour vision into account – i.e., to

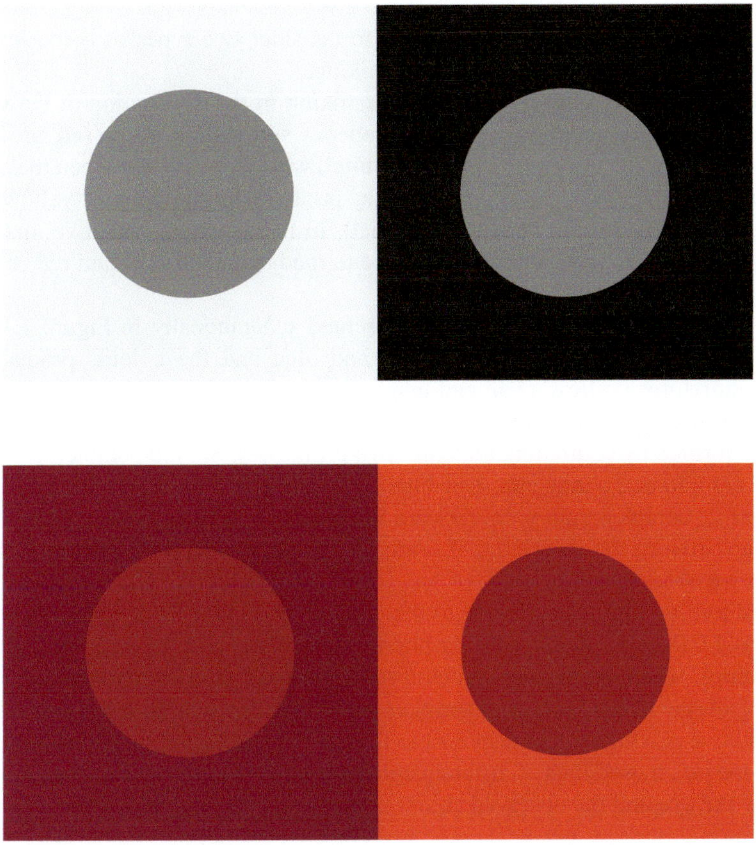

Figure 1.13 Examples of simultaneous contrast. In the top half of the figure, the apparent lightness of the identical grey discs is influenced by their background. In the lower half, the hue of the identical discs is perceived differently depending on the background colour

formulate a **colour appearance model (CAM)**. A detailed exploration of CAMs is beyond the scope of this book, but their development in the 1980s and 1990s is summarised by Fairchild (Fairchild 2005: 196–237). The Commission International de l'Éclairage (CIE)[3] published a colour appearance model (CIECAM97s) in the late 1990s. It was succeeded by the CIECAM02 model, which incorporated improved colour adaptation modelling (CIE 2004b) and the CIECAM16 model, which gives a better fit between theoretical and practical results, and is the current standard (CIE 2022c).

3 The CIE is the international standards body for matters of colour, light and lighting.

1.8 Additive and subtractive mixing

Before going further, it might be useful to consider two types of colour mixing, termed additive mixing and subtractive mixing.

We have already encountered additive mixing in the description of the way in which the eye senses white light. Simultaneous stimulation of the red, green and blue sensors produces the sensation of 'white', while we saw that when the signals from the red and green cones combine they produce the sensation of yellow (used in the eye's yellow-blue opponent channel). In the same way additive mixing of blue and green produces the greenish blue termed cyan and blue and red combine to give magenta (Figure 1.14 left).

In the model of **additive mixing** presented schematically in Figure 1.14, the three additive primaries are red, green and blue and the colours produced by their admixtures (yellow, cyan and magenta) are termed secondary colours. Each of the primaries can be paired with a complementary secondary colour (the secondary formed by combining the other two primaries). So, red complements cyan, green complements magenta, and blue complements yellow. As we will see in Chapter 2, complementary colours are important to help us understand the relationship between the absorption of some wavelengths of light by a material and the colour we observe.

Additive mixing is applicable to the creation of colour from emissive sources, such as LCD or LED displays on phones and other devices, and white fluorescent lamps, where phosphors that emit red, green and blue light are blended in the coating on the inside surface of the glass envelope, see Section 2.1.10.

Additive mixing is also responsible for the colour seen in pointillist paintings. Here, the juxtaposed dots of different colours are too small to be resolved individually by the eye and the observed colour is that of their additive mixture.

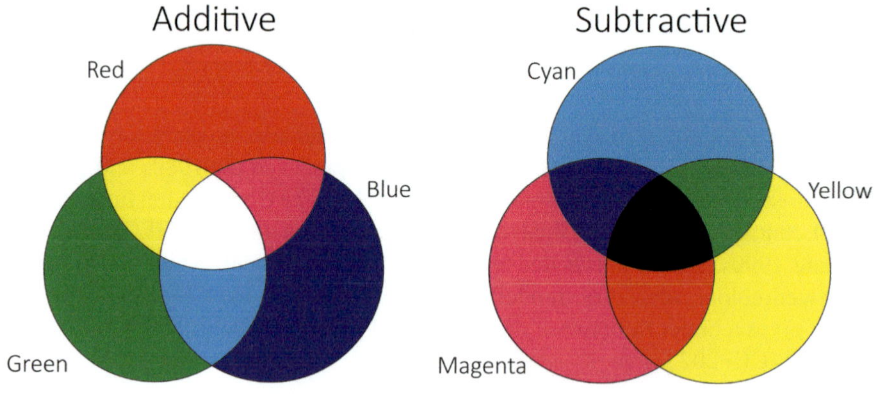

Figure 1.14 Schematic representation of additive and subtractive mixing

As the term suggests, subtractive mixing arises from the removal, by absorption, of certain wavelengths from the white light with which a material is lit. So, a material that absorbs red light will reflect light in the green and blue region, and the material will be coloured cyan. In the same way, the absorption of blue light will give a yellow colour and the absorption of green will give magenta. Although the subtractive primaries are sometimes termed blue, red and yellow, they are more correctly termed cyan, magenta and yellow, which might be familiar as the colours used in three-colour printing. When cyan (corresponding to the absorption of red light) and yellow (corresponding to the absorption of blue light) are combined, only green light is reflected, so this is the colour of the resulting mixture (Figure 1.14, right). In the same way mixtures of cyan and magenta produce blue and those of magenta and yellow yield red. As seen in Figure 1.14, a combination of all three subtractive primaries (cyan, magenta and yellow) absorbs all (or most) of the wavelengths from the incident white light and will appear black (or nearly black).

Subtractive mixing is responsible for the colours of the majority of painted or dyed objects, although artists have hardly ever used a simple set of three primaries. First, a white pigment is often needed to create lighter shades than those produced by combining two or more coloured pigments. Second, the accessibility of materials may affect choices so that, for example, it is easier to use a single readily-available pigment than a mixture of two 'primaries', although the opposite argument also frequently applies. Cost also plays a role in determining whether a particular material is employed rather than a mixture of cheaper pigments or dyes.

While blacks can be produced by combining other colours – a good example is the range of near blacks in the shadows of Monet's paintings of *Gare Saint Lazare* (Bomford et al. 1990: 168–169) – deeper (and cheaper) shades can be produced by the use or addition of a black pigment such as carbon black. These principles also apply to four-colour printing, where the three subtractive primaries (cyan, magenta and yellow) are supplemented by a black ink that can both give deeper blacks and reduce the consumption of the more expensive coloured inks.

1.9 Colour order systems

Since antiquity, we have sought to arrange or order colours, often linking them to elements of the world around us, our bodies, or our emotions. Aristotle, for example, associated earth, water, wind and fire to particular colours. These methods for arranging colours were frequently based on the primaries from either or both the additive and subtractive systems.

Newton's concepts of colour, published in *Opticks* in 1704, represented a radical departure from these earlier ideas in two respects. Not only was he the first to arrange colours in a now familiar circular pattern (Figure 1.15), but rather than drawing on a series of arbitrary primary colours, he placed the sequence of colours across the visible spectrum from violet to red around the circumference of his circle (Newton 1704).

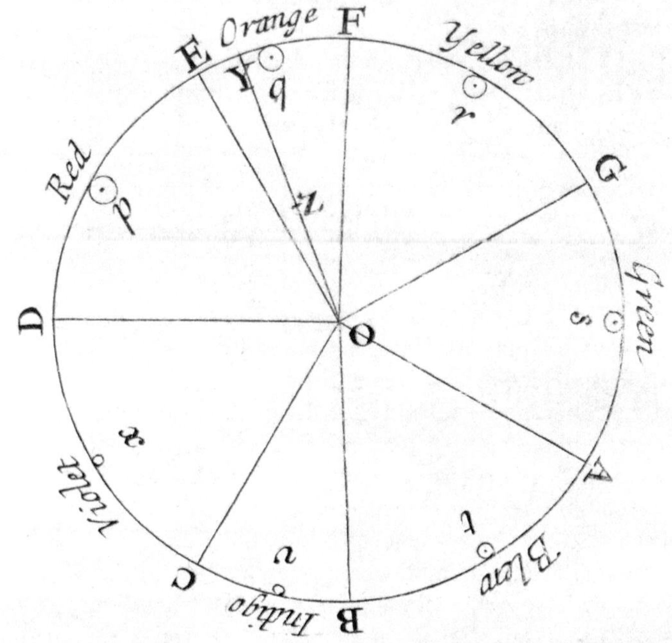

Figure 1.15 Colour circle from Newton's *Opticks* (1704)

Subsequently, every significant colour order system has adopted this circular model with the position around the circumference termed the **hue** of the colour. Systems published in the eighteenth and nineteenth centuries recognised that hue alone was not sufficient to define a colour, with many of the colour circles or wheels developed in that period also containing a scale of **lightness** or darkness to refine each hue, for example, Moses Harris's *The Natural System of Colours*, in which each hue is modified by the addition of black to produce darker shades (Figure 1.16).

Although other systems, notably Runge's *Farbenkugel* (colour spheres) presented colours in three dimensions, it was not, as we will see in Chapter 3, until the twentieth century that three parameters were defined to characterise colours and these were assigned numerical values. A history of the development of colour order systems is given by Kuehni and Schwarz (2008).

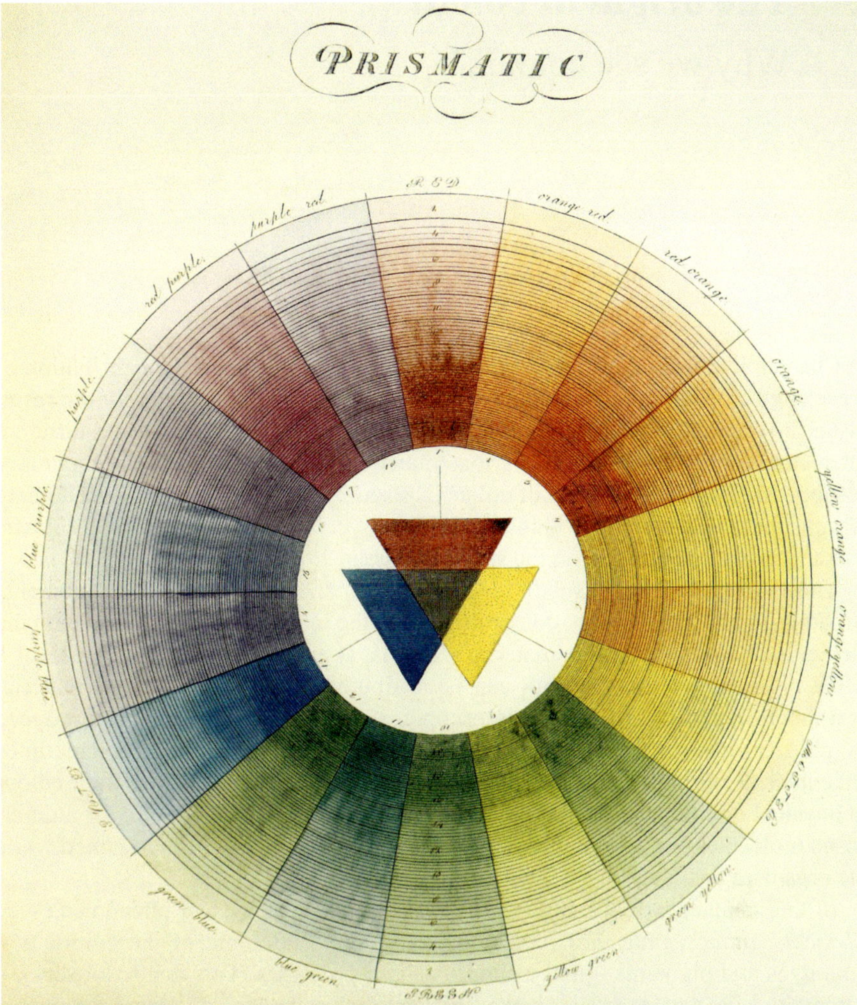

Figure 1.16 Colour circle from Moses Harris's *The Natural System of Colours* (1785)

2 The origin of colour

Why we see colours

In Chapter 1, we saw how our perception of colour arises from the distribution of wavelengths of light reaching the eye and that this, in turn, is generally the result of the interaction of white light with objects. But what lies behind the different interactions with white light of various materials? A well-known book on colour is subtitled 'The fifteen causes of color' (Nassau 2001), and there are indeed many reasons for the characteristic colours of objects, very readably set out in this classic text and in a shorter article on the subject (Nassau 1987).

These causes can be subdivided into chemical and physical phenomena. Broadly, chemical causes for colour arise from the interaction of radiation with the elements or compounds in a material – that is, at an atomic or molecular level – while physical causes of colour can be attributed to the structure of a material at a much larger, if still often microscopic, scale. As a result, colours produced by physical effects are often referred to as **structural colours**. We sometimes encounter the term 'biological pigment', but these materials also owe their colour to physical or chemical processes, albeit in sometimes complex plant- or animal-based molecules and with the caveat that the colour may only be maintained while the organism is alive.

In this chapter we will explore these chemical and physical phenomena (see *Elements, atoms, molecules and compounds* box) and some of the colorants – both dyes and pigments – that exemplify these processes. This is not, however, a comprehensive survey, which is well beyond the scope of this volume. Interested readers might refer to, *inter alia,* the *Pigment Compendium* (Eastaugh et al. 2004), the four volumes of *Artists' Pigments* (Feller 1986; Roy 1993; West FitzHugh 1997; Berrie 2007), *Natural Dyes* (Cardon 2007), *The Diversity of Dyes in History and Archaeology* (Kirby 2017) and the CAMEO online pigment database at MFA Boston (CAMEO n.d.).

DOI: 10.4324/9781003397182-3

Elements, atoms, molecules and compounds (see Book 1 in this series for more detail)

Elements are pure substances that cannot be decomposed into simpler substances by chemical processes (e.g., hydrogen or carbon). If a substance appears on the periodic table, it is an element.

Atoms are the smallest particles that retain the characteristic properties of a chemical element (e.g., a hydrogen atom or a carbon atom).

Molecules are electrically neutral entities comprising more than one atom. A molecule may contain atoms of the same element (e.g., a hydrogen molecule, H_2) or different elements (e.g., carbon dioxide, CO_2)

Compounds are chemical substances composed of many identical molecules, each of which contains two or more different elements (e.g., carbon dioxide, CO_2 or methane, CH_4).

2.1 Chemical causes of colour

2.1.1 Absorption

We saw in Chapter 1 (Figure 1.5) that black, white and grey arise when objects reflect all wavelengths of light more-or-less equally. Other colours arise from the selective absorption of some of the wavelengths present in white light. The colour we see, however, is not that of the light absorbed, but the complementary colour, as defined above. For example, the spectrum for red lead shown in Figure 1.5 indicates that the violet, blue and green portions of the spectrum are absorbed, with the result that the reflected light is richer in the orange and red wavelengths that impart the pigment's colour. Absorption of radiation of a particular wavelength occurs when its energy is equal to that of an energy level transition in the atom or molecule. Many of these transitions will have energies that correspond to wavelengths outside the visible region of the spectrum, so will not give rise to colour.

2.1.1.1 Atoms and absorption

The behaviour of atoms offers a relatively simple starting point for considering the absorption of radiation, even if few colorants are based on elements rather than compounds. Even so, the explanations given here are a simplified account, covering only the basics necessary to understand the origins of colour.

The classical model for an atom is of a nucleus containing protons and – except in the case of hydrogen – neutrons. The nucleus is surrounded by as many electrons as there are protons (to maintain electrical neutrality, as protons and electrons have positive and negative charges, respectively). The simplest atom is that of hydrogen, with a single proton and electron.

In the description of atoms in Book 1 of this series, these electrons were found in a series of 'shells' surrounding the nucleus and, where there was more than one electron present, tended to exist in pairs. We can, however, define these shells in a little more detail, as each comprises so-called 'orbitals', each of which can accommodate two electrons. These orbitals are of four different types, termed 's', 'p', 'd' and 'f', which have different shapes. The first shell has only a single s orbital, which we give the designation '1s'. The second shell contains s and p orbitals, termed 2s and 2p respectively. There is one 2s orbital, but three 2p orbitals, each with the same shape, but oriented differently, so there are eight electrons in this shell, as described in Book 1.

As we move through the periodic table, the number of protons and electrons increases in tandem (to maintain electrical neutrality) and defines the atomic number of the element – for example, sodium has atomic number 11, as it contains 11 protons and 11 electrons. As the number of electrons increases, they fill the atomic orbitals beginning with that of lowest energy, following the so-called **Aufbau principle.** Figure 2.1 shows the relative energy of the first ten levels; the 1s, 2s, 2p, 3s, 3p, 4s, 3d, 4p, 5s and 4d. We can see that although there are three p orbitals in any given shell, they each have the same energy, as do the five 3d and 4d orbitals in the third and fourth shells (Figure 2.1). This shows only the electronic levels occupied in elements with atomic number up to 48 (cadmium). The electrons in heavier elements fill further s, p and d orbitals as well as the fourth class of orbital, the f orbital.

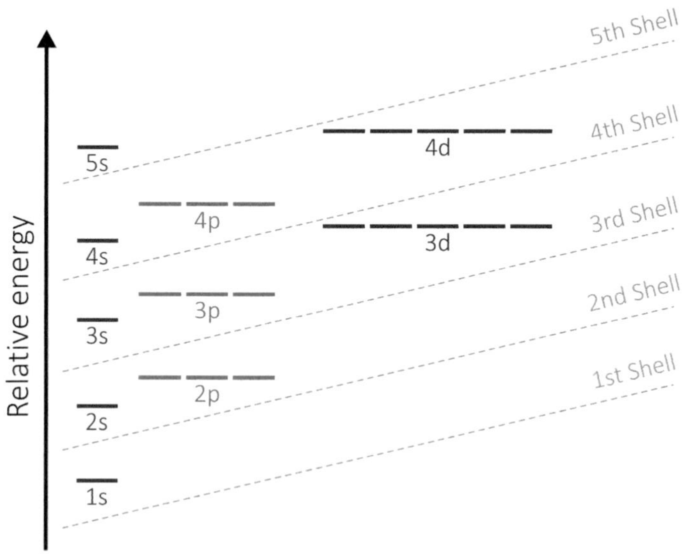

Figure 2.1 Schematic representation of the relative energy of the first ten atomic orbitals

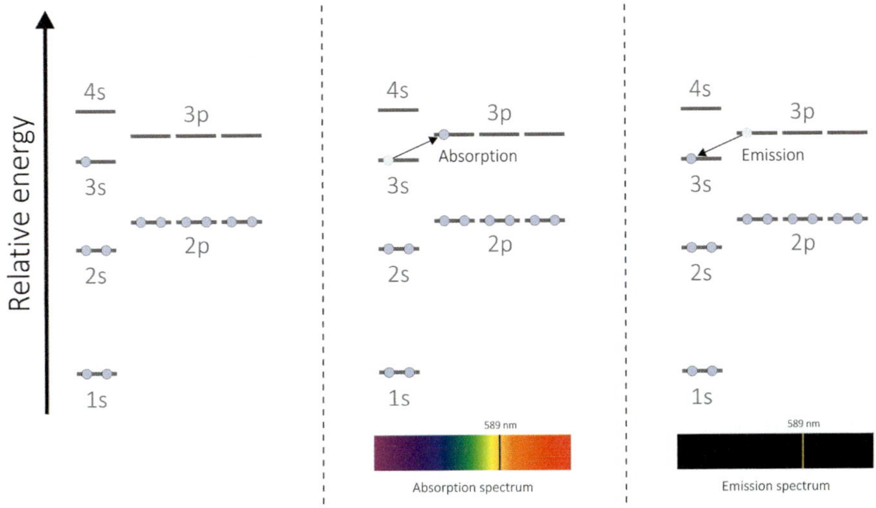

Figure 2.2 Schematic representation of the electronic configuration of sodium (left) showing the absorption due to the excitation of an electron from the 3s to 3p orbital (centre) and the emission due to the transition of an electron from 3p to 3s (right)

Returning to the example of sodium, the 11 electrons will fill the lowest energy levels available – two in the 1s orbital, two in the 2s orbital, two in each of the three 2p orbitals and one in the 3s orbital (Figure 2.2 left). This electronic configuration, written as $1s^2 2s^2 2p^6 3s^1$, represents the lowest energy state or 'ground' state of the atom.

An extremely common route for the absorption of energy is through the excitation of an electron from the highest occupied level – in this case, the 3s orbital – to the lowest unoccupied level, here the 3p orbital. The energy difference between these two levels is 2.1 eV (see *Energy levels and the absorption of radiation* box), which corresponds to radiation with a wavelength of *c*.589 nm in the visible region.[1] If we pass white light through sodium vapour and then split the radiation using a prism, we observe a dark line in the spectrum at 589 nm as this radiation has been absorbed by the sodium atoms and is absent from the spectrum (Figure 2.2 centre).

If, on the other hand, we heat or pass an electrical current through sodium vapour, then the electron in the 3s orbital may be excited to the 3p electronic orbital. When this electron then returns to the ground or 3s level, radiation with a wavelength of 589 nm will be emitted, giving the characteristic yellow colour of the sodium

1 The line at *c*.589 nm is technically a double line at 589.0 and 589.6 nm, caused by spin-orbit splitting of the 3p energy level. The difference in the energy of the two lines represents the difference between the $^1/_2$ and $^3/_2$ spin states.

lamps once commonly found in streetlights (Figure 2.2 right). Radiation will also be emitted as electrons excited to higher levels return to the ground state but as this radiation is of higher energy, it will be outside the visible spectrum (for example, the energy transition from 4p to 3s corresponds to a wavelength of *c*.330 nm, which is in the ultraviolet region).

Energy levels and the absorption of radiation

The difference in energy between two electronic levels in an atom or molecules is generally given using the unit of electron volts (eV).

Energy is related to wavelength by the relationship given in Chapter 1 ($E = hc/\lambda$), so we can calculate the range of energy differences that will give rise to absorptions in the visible spectrum and may thus impart colour to materials. The visible region from 400–700 nm equates to energies of between 3.1 and 1.7 eV.

Electronic transitions greater than 3.1 eV correspond to radiation in the ultraviolet range and beyond, while energies less than 1.7 eV fall in the infrared region; neither will give rise to colour.

There are, however, rather few instances in which colour in cultural heritage objects is the result of emission from atoms. Notable exceptions are in neon sculptures and signs, some of which are of considerable historical significance. The electrical excitation of a tube containing neon gas causes a characteristic red emission at *c*.640nm when the electrons return to the ground state. Gas-filled tubes that emit other colours are often termed 'neon lights' but either contain other gases – for example, argon produces a blue-purple light – or, more frequently, have a phosphor coating on the inner surface of the glass tube that produces coloured light by luminescence (Figure 2.3), a process covered in Section 2.1.10 on photoluminescence.

Before moving on, it is worth noting that not all transitions between different energy levels are possible. A set of so-called selection rules, the details of which are beyond this simple introduction, dictate which transitions are 'allowed' or 'disallowed' in atoms and molecules. In practice, these disallowed transitions sometimes occur, but tend to give rise to weaker absorptions than those that are allowed.

2.1.1.2 Molecules and absorption

The situation for compounds and molecules is both more complicated and more diverse. When elements combine to form compounds, the atomic orbitals of the component atoms also combine to form molecular orbitals and it is the transitions between these molecular orbitals that give rise to the absorptions that produce the

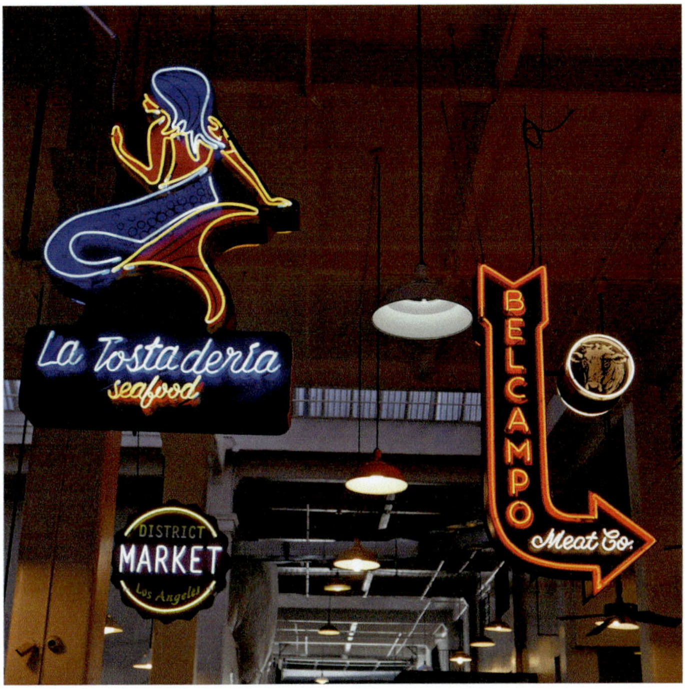

Figure 2.3 Light art in Central Market, Los Angeles

colours of these substances. A thorough description of molecular orbital theory is beyond this survey of the colours to which they give rise, but we can look at some simple examples that help to explain the basis of chemical bonds and will serve as a useful background when we come to look at larger molecules.

Although not a colorant, one of the simplest compounds is lithium hydride (LiH).[2] As we have seen, hydrogen is the simplest atom, with electronic configuration $1s^1$, while lithium – the third element in the periodic table – can be represented as $1s^2 2s^1$. The energy levels of lithium (Li) and hydrogen (H) are shown on the left in Figure 2.4. One way of considering the interaction between Li and H is to posit the formation of a bond between these atoms. In this model of bonding, the 2s *atomic* orbital of Li and the 1s *atomic* orbital of H combine to form two new *molecular* orbitals, seen on the right in Figure 2.4. The type of bond formed between Li and H

2 For the purposes of this discussion of molecular orbitals, LiH is considered as a simple binary compound with a bond between the component atoms. In reality, LiH exists in a polyatomic lattice form, with each lithium or hydrogen atom associated with its six nearest neighbours in a crystalline network.

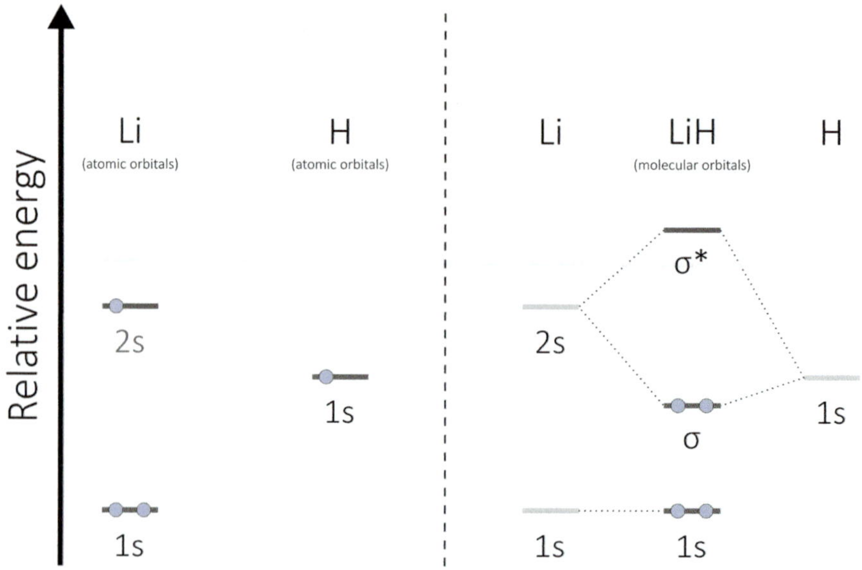

Figure 2.4 The occupied atomic orbitals in lithium and hydrogen (left) and the molecular orbitals created when they form lithium hydride (right)

is termed a sigma (denoted by the Greek letter σ) bond and the two new associated molecular orbitals are labelled σ and σ*. The σ and σ*molecular orbitals are termed the σ bonding and σ antibonding orbitals respectively. Simplistically, occupancy of bonding orbitals by electrons tends to strengthen a bond, while electron occupancy in antibonding orbitals weakens a bond. As can be seen in Figure 2.4, two of the electrons in LiH remain in the lowest energy level 1s orbital of Li while the remaining electrons now occupy the σ orbital, promoting bonding between the lithium and hydrogen atoms.

Figure 2.4 can also be used to illustrate the concepts of the highest occupied molecular orbital (HOMO) and the lowest unoccupied molecular orbital (LUMO). Highest and lowest here refer to the relative energies of the orbitals, so in LiH the HOMO is the σ molecular orbital, occupied by two electrons, while the LUMO is the σ* molecular orbital. As it is between the HOMO and LUMO that most transitions occur, the energy difference between them determines the wavelengths of radiation most strongly absorbed and hence the colour (if any) of the compound. As the energy difference between the HOMO and LUMO in LiH equates to an absorption in the ultraviolet region, the material is not coloured.

2.1.1.3 Vibrational levels

One of the first features that becomes apparent when we look at most absorbance spectra for compounds – for example, in Figure 1.5 or Figure 2.17 later – is that

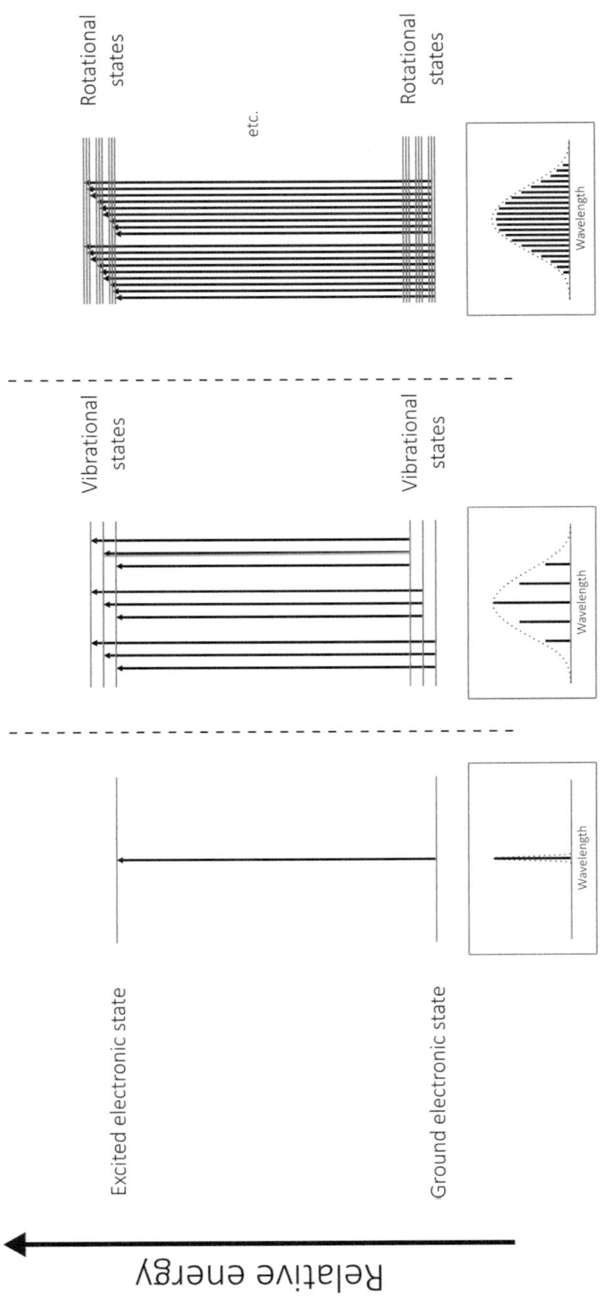

Figure 2.5 The broadening of electronic absorption peaks caused by excitation to different vibrational and rotational states

they do not contain the narrow bands of the atomic spectra of sodium or neon. This is because for each of the electronic energy levels shown in Figure 2.4 there are a number of sub-levels, termed **vibrational energy levels** as they arise from vibration of the bonds that join atoms in molecules. Figure 2.5 shows a common method of representing these vibrational levels within each electronic level. We can see that if absorptions occur from several different vibrational levels in the lowest (ground) energy level to vibrational levels in the excited energy level, radiation of a number of wavelengths will be absorbed. Statistics dictate occupation of these levels and there are rules about transitions, but we generally expect a symmetrical Gaussian distribution of absorptions centred on the wavelength associated with the transition from equivalent vibrational energy levels in the ground and excited electronic states – see Figure 2.5 centre, for a much-simplified representation.

Transitions are possible from one vibrational level to another in the same electronic level, but the energy difference between these will be much smaller and the radiation that is absorbed will lie in the infrared region of the electromagnetic spectrum. While these absorptions will not contribute to the colour of materials, they are helpful in material identification using infrared spectrometry.

2.1.1.4 Rotational levels

In the same way that each electronic level may contain many vibrational levels, so the latter are subdivided into a number of **rotational energy levels**, each associated with the different rotational states that exist in all but the very simplest molecules, Figure 2.5 right. The existence of multiple rotational energy levels further contributes to the broadening of absorption spectra. As the energy differences between rotational levels are even smaller, transitions between them again equate to wavelengths well outside the visible region, meaning that they play no part in determining the colour of objects, but can again be useful in spectroscopic analysis of materials.

2.1.2 Transition metal colours

Many of the colours that we encounter in pigments, glasses and ceramic glazes are the result of selective absorption by metals in the first row of transition metals in the periodic table.[3] From an electronic perspective, these metals are characterised by the presence of electrons in the 3d orbitals of their atoms and ions, which are active in bonding with other elements. As we saw earlier (Figure 2.1), there are five 3d orbitals each capable of holding two electrons. In general, colours arise in those ions that do not have a full complement of ten electrons – so compounds based on

3 The first row of transition metals comprises scandium (Sc), titanium (Ti), vanadium (V), chromium (Cr), manganese (Mn), iron (Fe), cobalt (Co), nickel (Ni), copper (Cu) and zinc (Zn).

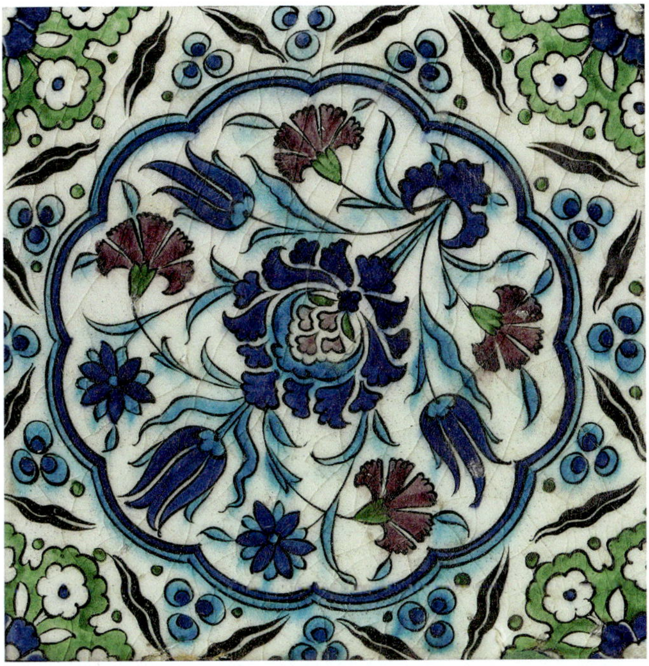

Figure 2.6 Seventeenth-century Iznik ceramic tile with glazes based on manganese (purple), cobalt (blue) and copper (turquoise) [Source: Turkey, 1600s, Ottoman Period. 26.8 x 26.8 cm. The Cleveland Museum of Art, 1915.651.c. Gift of the John Huntington Art and Polytechnic Trust.]

the copper ion Cu^{2+}, which has configuration $3d^9$, tend to be coloured while those of the zinc ion Zn^{2+} ($3d^{10}$) do not, and appear white.

In the first-row transition metal elements the five 3d orbitals all have the same energy but, because these orbitals have different shapes, when they form compounds, there is a 'split' in the energy levels caused by different degrees of interaction with surrounding atoms or groups of atoms. As a result, some of the 3d orbitals will have higher energy than others and transitions (termed d–d transitions) can occur between them through the absorption of radiation. In many compounds this energy difference equates to a wavelength in the visible spectrum, creating characteristic colours. For example, glasses and ceramic glazes have long been coloured with compounds of chromium (Cr^{3+}: green), manganese (Mn^{2+}: purple), iron (Fe^{2+}: yellow to green), cobalt (Co^{2+}: blue), nickel (Ni^{2+}: brown) and copper (Cu^{2+}: turquoise to green); the tile illustrated in Figure 2.6 owes its colours to manganese (purple), cobalt (blue) and copper (turquoise). Early 'colourless' glass often appears green, because iron is such a frequent contaminant in its raw ingredients.

A complete description of why and how the 3d orbitals split into groups with different energies is beyond this book, but is related to the geometry of the groups of atoms (often called ligands in this context) that surround the metal ion, the nature of these ligands and the **oxidation state** of the metal (see *Ions and oxidation state* box).

Ions and oxidation state

Atoms and molecules are electrically neutral entities in which the number of positively charged protons exactly matches the number of negatively charged electrons. In ions, the number of electrons and protons are not equal. Positive ions contain fewer electrons than protons while negative ions contain more electrons than protons.

For example, if a single electron is lost from an atom of sodium (Na), a sodium ion with a single positive charge is created, written as Na^+. Likewise, if an oxygen atom gains two electrons, it becomes a doubly negative ion (O^{2-}).

The number of electrons lost is sometimes referred to as the oxidation state of the ion, so copper that has lost two electrons is said to have an oxidation state of plus two (+2), this is generally written as copper(II) when a compound is named in full, or Cu^{2+} when referring to the ion. In a similar manner, a chlorine atom that has gained an electron to become a chloride ion has an oxidation state of minus one (-1), written as Cl^-. In some instances, particularly in complicated structural formulae, the oxidation state is given by a superscript Roman numeral, for example, Cu^{II}.

In the same way, molecules can also gain or lose electrons to become negative or positive ions.

Two common arrangements of ligands around transition metal ions are so-called octahedral and tetrahedral symmetries. The relationship between the ligands and the five 3d orbitals is different in each case and causes a different pattern of splitting of the energies of the orbitals (Figure 2.7). We can see that the energy difference between the two groups of orbitals is greater for ligands in an octahedral arrangement than for a tetrahedral configuration.

The energy difference for a particular symmetry is increased if the oxidation state of the transition metal is increased. Although certain ligands produce a greater energy difference than others, they may also change the symmetry, so it is difficult to disentangle these connected effects on the colour produced by the metal ion. In some instances, changing the ligands can distort the regular tetrahedral or octahedral symmetry,[4] creating additional energy levels and, hence, altering the colour.

4 The distortion of octahedral geometry in which the two opposite ligands move away from the metal ion is quite common and is termed the Jahn Teller effect.

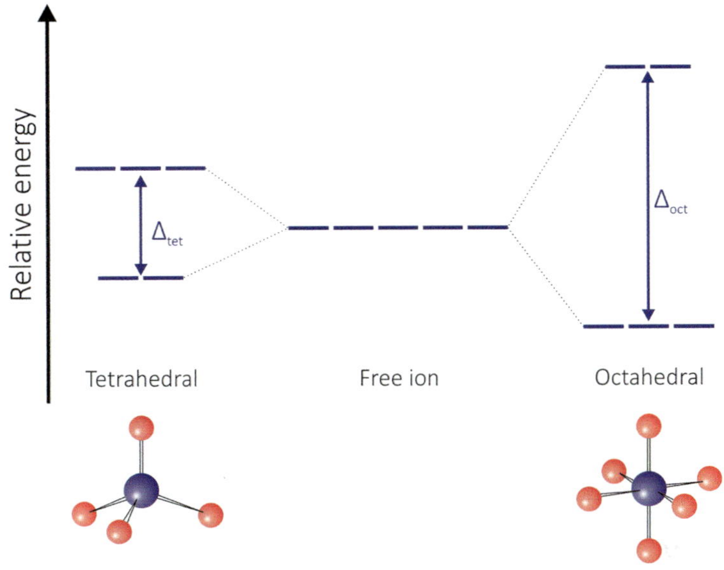

Figure 2.7 Splitting of the energies of the five d orbitals in transition metals by an octahedral or tetrahedral arrangement of ligands around the metal ion

An example is provided by the copper carbonate pigments azurite and malachite which have been widely used for many centuries and by many cultures, either derived from natural minerals or manufactured from other copper compounds, see Figure 2.8. Although both malachite and azurite contain the same ions, copper (Cu^{2+}), carbonate (CO_3^{2-}) and hydroxide (OH^-), their proportions differ – malachite is $Cu_2(CO_3)(OH)_2$ and azurite is $Cu_3(CO_3)_2(OH)_2$. As a result, the arrangement of ligands around the copper ions also differs, causing a change in the energy levels and altering the wavelengths of light that are absorbed, yielding a blue colour for azurite and green for malachite.

Copper-containing greens are not limited to malachite, with several other minerals used as colorants, including various basic copper chlorides (i.e., compounds containing both hydroxide and chloride ions) and basic copper sulfates such as posnjakite ($Cu_4(SO_4)(OH)_6{\cdot}H_2O$). Synthetic copper greens have also been important colorants from antiquity. Perhaps the most significant is verdigris, basic copper acetate – $Cu_4(O_2CCH_3)_2(OH)_6$, manufactured by exposing copper metal to acidic fumes from, for example, vinegar or pomace (the residues left from winemaking or cidermaking).

Two highly toxic arsenic-containing greens manufactured in the late eighteenth and early nineteenth century are also based on copper. Scheele's green (copper arsenite: $Cu(O_2AsOH)$) dates from the 1770s, while the more vivid emerald green (copper acetoarsenite: $Cu_4(O_2CCH_3)_2(AsO_2)_2$), which is sometimes known as Schweinfurt, Vienna or Paris green, was introduced in the 1810s (CAMEO n.d.).

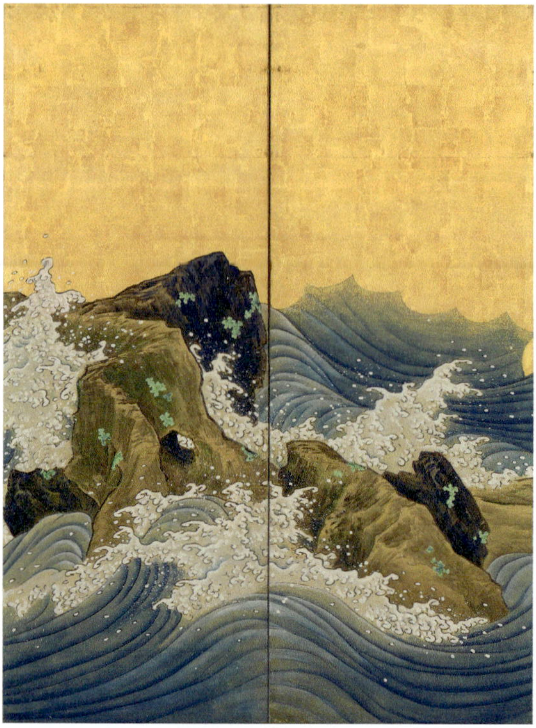

Figure 2.8 Two panels from a Japanese folding screen *Rocks amid Crashing Waves (Hatōzu)* showing the use of blue (azurite) and green (malachite) copper-based pigments [Source: Yamaato Kakurei, *c.*1810. *c.*165 x 486 cm. Los Angeles County Museum of Art, M.2017.90. Gift of the 2017 Collectors Committee with additional funds provided by Laurie and Bill Benenson.]

Other blue copper-based colorants include Egyptian blue ($CaCuSi_4O_{10}$) and Han blue ($BaCuSi_4O_{10}$), which although structurally similar (differing only in the substitution of barium for calcium in the latter), were developed – as their names suggest – in widely different cultural contexts. Unlike azurite, both these pigments were manufactured rather than relying on the processing of naturally occurring minerals. While Han blue has been reported predominantly in Chinese objects, the use of the very stable Egyptian blue spread beyond the Nile Valley and is found in Greek, Roman and later European contexts (Figure 2.9).

Han purple ($BaCuSi_2O_6$), also developed in China at least two millennia ago (West FitzHugh and Zycherman 1992), is structurally very similar to Han blue, again highlighting how small changes in the ligands surrounding a metal centre can affect colour. We will revisit the importance of the both the number and nature of the ligands in Chapter 5, when we come to consider the way in which transition metal-based colours can alter in response to a change in the ligands surrounding the metal.

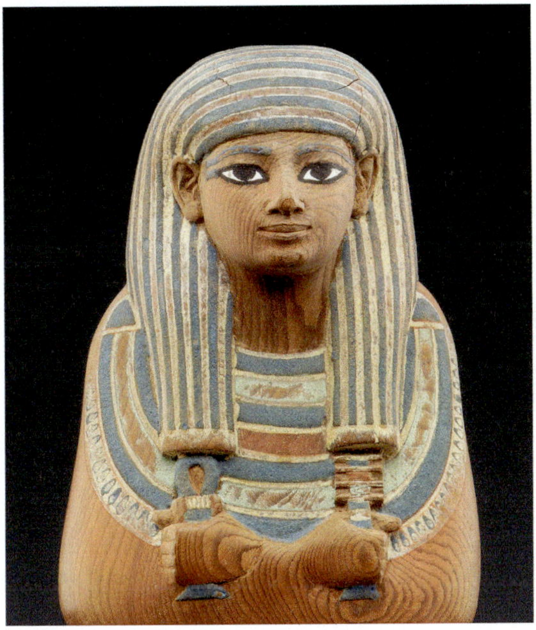

Figure 2.9 Carved cedar wood shabti of Yuya decorated with Egyptian blue pigment [Source: Egyptian, from the tomb of Yuya and Tjuyu at Thebes, New Kingdom, 18th Dynasty (c.1390–1352 BCE). c.28.5 x 7.8 x 5.4 cm. Metropolitan Museum of Art, 30.8.57. Theodore M. Davis Collection, Bequest of Theodore M. Davis, 1915.]

Cobalt-containing compounds are also an important source of blue colorants and cobalt-bearing ores have been processed and refined for use in ceramic glazes, including in the famous blue and white ceramics from China, later European reproductions and near Eastern ceramics such as that shown in Figure 2.6. Analogous processes were employed to manufacture blue glass for vessels and stained glass windows (Figure 2.10), and to produce the blue pigment smalt, which is essentially a ground cobalt-containing glass (see Chapter 5 for more details about smalt and its deterioration).

Research into ceramics, glasses and pigments has shown that the blue colour originates from absorptions by tetrahedrally co-ordinated cobalt(II) ions, although the nature of the bulk glaze or glass and any impurities carried through from the cobalt ore will influence the colour. These impurities are, incidentally, often useful in suggesting a possible geographic origin for the cobalt ore.

The nineteenth century saw the synthesis and introduction of a number of cobalt-based blue and purple pigments. The earliest of these, cobalt blue or Thénard's blue, is a cobalt aluminate ($CoAl_2O_4$), which is broadly similar to the less pure materials used in the production of blue glazes on ceramics. Other

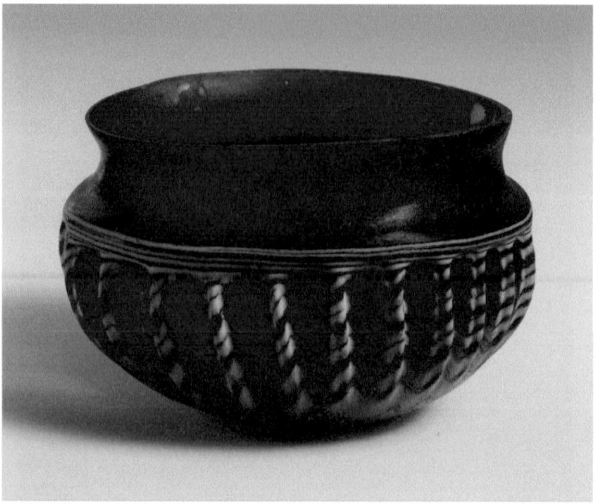

Figure 2.10 Ribbed glass bowl that owes its blue colour to the addition of cobalt to the glass. [Source: Roman, mid-first century CE. *c*.8.3 Ø x 6.2 cm. Metropolitan Museum of Art, 91.1.1247. Edward C. Moore Collection, Bequest of Edward C. Moore, 1891.]

nineteenth-century introductions included violet pigments based on either cobalt phosphate ($Co_3(PO_4)_2$) or cobalt arsenate ($Co_3(AsO_4)_2$) and the blue-green pigment cerulean blue (cobalt stannate: Co_2SnO_4).

In contrast, the pigment aureolin (potassium cobalt(III) nitrate: $K_3[Co(NO_2)_6]$), which was introduced in the second quarter of the nineteenth century, contains Co^{3+} ions surrounded by ligands in an octahedral arrangement. This combination of symmetry, ligands and oxidation state gives rise to a different pattern of absorption, with the result that aureolin is a yellow pigment.

Green pigments based on chromium(III) were introduced in the nineteenth century, principally the chromium oxide, chrome green (Cr_2O_3), or the brighter hydrated oxide, viridian ($Cr_2O_3{\cdot}xH_2O$). The yellow pigments lead chromate (chrome yellow: $PbCrO_4$) and zinc chromate (zinc yellow: $ZnCrO_4$) also owe their colour to chromium, but as chromium(VI), illustrating the importance of oxidation state in determining colour, see Section 2.1 and Chapter 4, Section 4.1.3.2.1.

2.1.3 Colour from f–f transitions

As mentioned earlier, heavier elements possess so-called f orbitals in addition to the s, p and d orbitals illustrated in Figure 2.1. The elements for which the seven 4f orbitals (each capable of holding two electrons) are the highest energy orbitals that are occupied are termed the lanthanides (named after the lightest element in the series – lanthanum) while those based on the 5f orbitals are the actinides (named after the element actinium). Those compounds containing a lanthanide or actinide

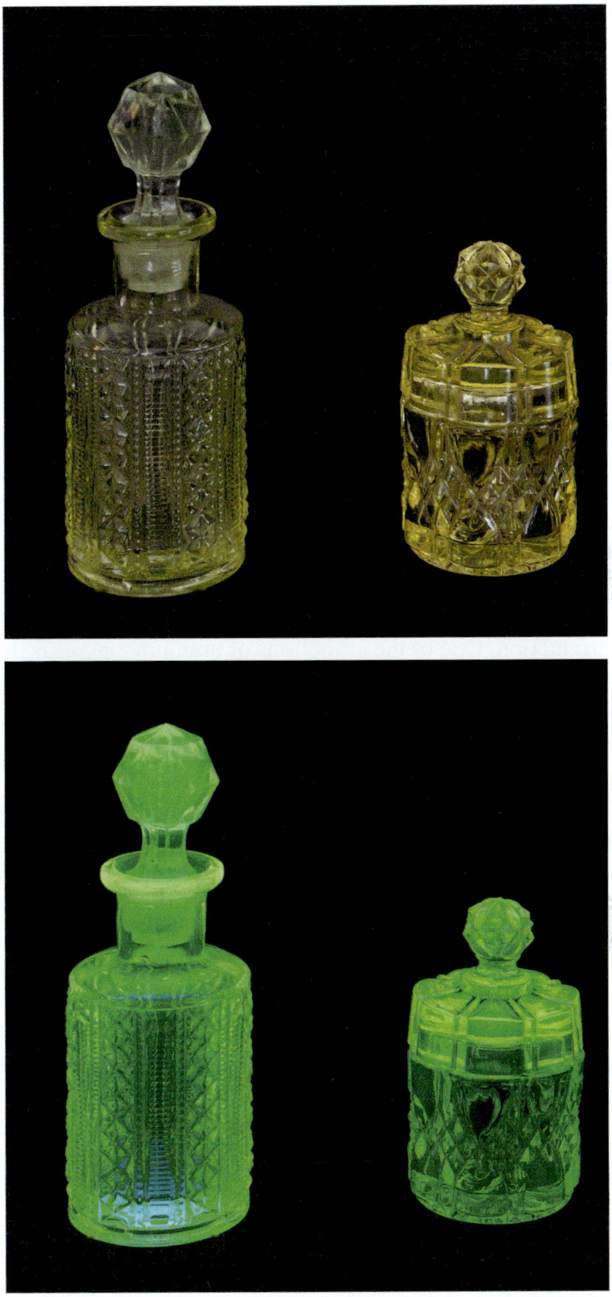

Figure 2.11 Glass bottles made of uranium glass illuminated with normal light (top) or ultraviolet radiation (bottom) [Source: © Raimond Spekking / CC BY-SA 4.0 (via Wikimedia Commons) (https://commons.wikimedia.org/wiki/File:Glass_flacon_under_UV_light-1393.jpg) and (https://commons.wikimedia.org/wiki/File:Glass_flacon_under_normal_light-1394.jpg).]

that are coloured owe this to absorption of radiation to promote f–f transitions. In both series the colour is largely dependent on electronic configuration, with the surrounding atoms or ligands exerting less effect than in some transition metal compounds.

The lanthanides are not common colorants, but cerium(IV) oxide (CeO_2) and praseodymium(III) oxide (Pr_2O_3) have been used to create brown and green glasses respectively, while neodymium(III) oxide (Nd_2O_3) gives a lilac to pink colour in glasses ('Alexandrite glass') and ceramics.

The pattern of colour in the actinides is similar, with colour related principally to electronic configuration. Although the 5f–5f transitions in actinides generally give rise to stronger colours than the 4f–4f transitions in the lanthanides, the former are less frequently used as colorants, perhaps because just the first four of these elements were known before 1940 and of those only the heaviest, uranium, produces coloured compounds. Ions of the actinides discovered subsequently (the transuranic elements) can produce colour, but they are scarce and (often highly) radioactive.

Glasses containing uranium, particularly those based on the UO_2^{2+} ion of uranium(VI), are the most common example of colour deriving from f–f transitions. These yellow glasses were introduced in the early nineteenth century and continued to be made until the second half of the twentieth century (Figure 2.11), when concerns about their radioactivity and the availability of cheaper alternatives ended commercial production. One of the attractions of uranium glasses is that in addition to their yellow body colour, they show a green fluorescence – discussed in more detail later in this chapter.

2.1.4 Charge transfer colours

The electronic transitions described in the previous two sections involve energy levels associated with the metal ion in a compound, although we have seen that the nature of the transition, and hence the energy of radiation absorbed, are influenced by the surrounding ligands.

In some cases, however, the absorption of energy causes an electron to be excited from an energy level associated with one part of a molecule to that associated with another part of the same molecule. As the electrons involved in these transitions bear a negative charge, this process is termed a **charge transfer** (between the two parts of the molecule).

Common examples include electronic transitions from an electronic level of a metal ion to a level associated with its ligand(s), termed metal–ligand charge transfer (MLCT) or *vice versa*, ligand–metal charge transfer (LMCT). While there are examples of MLCT in the literature, these do not include any obvious colorants found in the cultural heritage field, where we more frequently encounter transitions of the ligand–metal type.

A number of transition metal oxides owe their colour to LMCT, with electrons from the oxygen excited to a d orbital on the metal. Examples include the various yellow and red earth pigments based on iron oxides; massicot (lead oxide: PbO); lead chromate (chrome yellow: $PbCrO_4$); and lead antimonate (Naples yellow: $Pb_3(SbO_4)_2$).

In addition, some sulfides used as pigments – including vermilion (mercury sulfide: HgS) and cadmium yellow (cadmium sulfide: CdS) – as well as the less frequently encountered iodine yellow (lead iodide; PbI_2) and iodine red (mercury iodide; HgI_2) can also be considered as LMCT colours, although we will see later that these oxides, sulfides and iodides are sometimes also classified as band gap colours.

A type of charge transfer with greater relevance is termed intervalence charge transfer (IVCT), in which an electron (a charge) is transferred between two atoms with different oxidation states. Charge transfer between lead atoms in different oxidation states is responsible for the strong red colour in the pigment red lead (Pb_3O_4), which contains both Pb^{2+} and Pb^{4+} ions and can also be written as $Pb^{II}_2Pb^{IV}O_4$, while

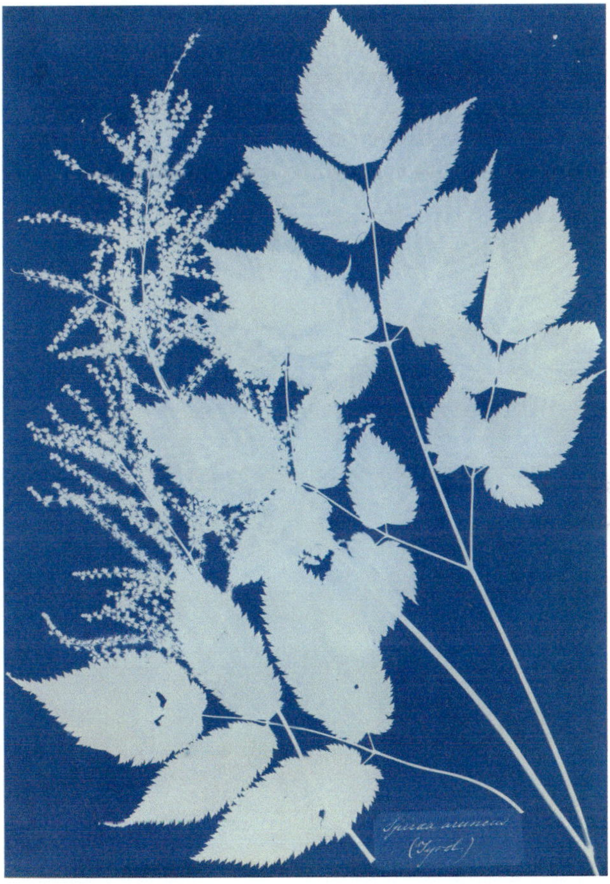

Figure 2.12 Cyanotype of *Ptilota plumosa* from a collection depicting British algae [Source: Anna Atkins, 1853. New York Public Library, Spencer Coll. Eng. 1843 93-440.]

IVCT between iron as Fe^{2+} and Fe^{3+} contributes to the characteristic colour of green earth pigments.

Perhaps the most notable example of a material that owes its colour to IVCT between metal ions is Prussian blue. Prussian blue and analogous compounds are used in pigments and inks, and are responsible for the blue colour of cyanotypes (Figure 2.12). Although there are a number of slightly different chemical compounds in the Prussian blue 'family', all contain iron in both the Fe^{2+} and Fe^{3+} oxidation states, surrounded by six cyanide ligands; one of the simplest members of this family has the formula $KFe^{III}[Fe^{II}(CN)_6]$. The absorption of red radiation centred at 680 nm, which leads to the transfer of an electron from Fe^{2+} to Fe^{3+} yielding $KFe^{II}[Fe^{III}(CN)_6]$, is responsible for the pigment's blue colour. In common with other charge transfer colours, the absorption of radiation is very strong, producing an intense deep blue pigment. This is because, unlike transitions between some of the electronic levels within metal ions, charge transfer transitions are allowed under the selection rules briefly mentioned earlier.

This strength of colour, and its relatively low cost, meant that after its (accidental) synthesis in 1704, Prussian blue was quickly recognised as a valuable addition to the artist's palette. It offered an alternative to established blue pigments that were either prone to change colour (indigo and smalt) or extremely expensive, in the case of ultramarine, whose sole source for many centuries was in Badakhshan in Central Asia.

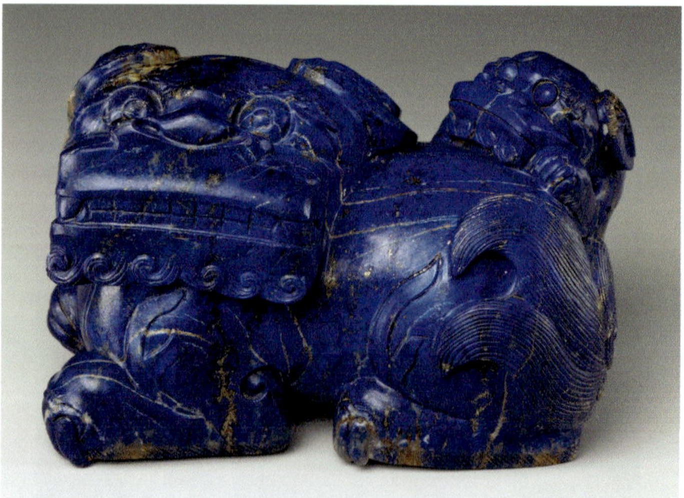

Figure 2.13 Chinese lion and cub sculpture carved from lapis lazuli [Source: Qing dynasty (1644–1911). 9.8 x 6 cm. Metropolitan Museum of Art, 02.18.922. Gift of Heber R. Bishop, 1902.]

The deep blue (gem)stone lapis lazuli, which is veined with quartz and pyrite, is used in its own right (Figure 2.13), but was also processed to extract the component blue mineral lazurite (the pigment ultramarine). Lazurite – and the manufactured 'artificial' ultramarine that is now used more widely – are sodium aluminosilicates (with calcium sometimes present in the mineral form). The aluminosilicate lattice holds and stabilises polysulfide ions (e.g., S_3^-), from which the colour derives, due to charge transfer between sulfur atoms in different oxidation states.

2.1.5 The colour of metals

We saw earlier that when two atoms combine, new energy levels are created, with energies different to those in the atoms themselves. If a greater number of atoms combine, the number of energy levels created will increase further, potentially creating a 'band' of energy levels. Metals provide a good illustration of how such bands can be produced. In individual metal atoms, all the energy levels will be identical – including those that will be involved in bonding (Figure 2.14). When two atoms bond, two energy levels are created – the bonding and antibonding levels that we saw earlier. If four atoms are involved in bonding, four energy levels will be created, and so on until – in an ordered structure such as a metal or a crystalline solid – there are so many levels of similar energy that they effectively form an **energy band** (Figure 2.14). In metals, this band, termed the conduction band, is responsible for many of their properties, not least their electrical conductivity which stems from the ability of these so-called delocalised electrons in the

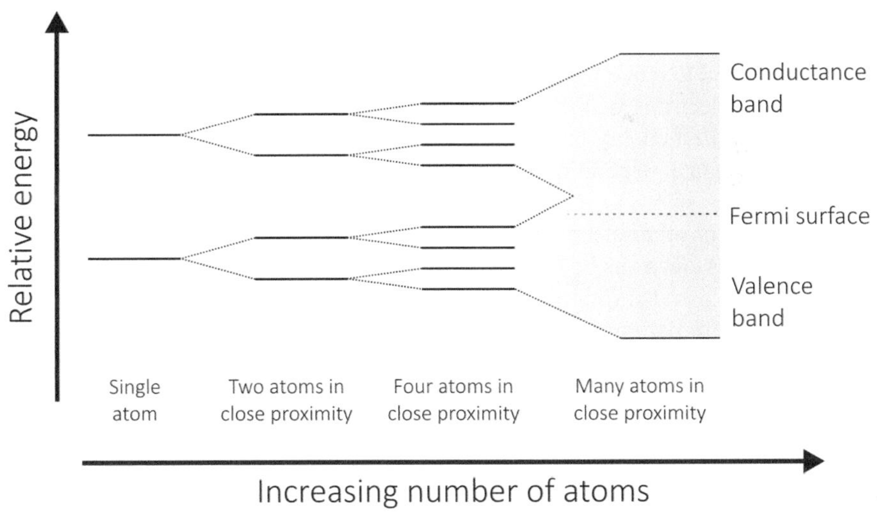

Figure 2.14 Schematic representation of the formation of electronic bands as the number of atoms in close proximity increases

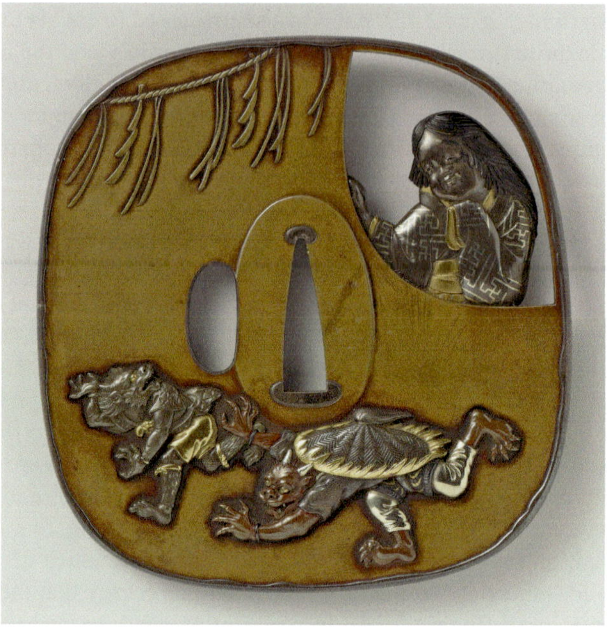

Figure 2.15 Japanese *tsuba* (hand guard) made from copper alloy, with decorations in gold, silver and other copper alloys [Source: Hagiya Katsuhira, 1825–1875. 9 × 8.7 cm. Rijksmuseum, AK-MAK-1097. On loan from the Royal Asian Art Society in The Netherlands (Westendorp-Osieck bequest, 1968).]

conduction band to move freely through the material. Within this conduction band, the available electrons will fill the energy levels, starting with the lowest energy levels. The highest energy level within the conduction band that is filled will vary from metal to metal depending on its electronic structure and is termed the Fermi surface.

When light or other electromagnetic radiation is absorbed at the surface of the metal, it excites electrons below the Fermi surface into higher energy levels in the band. Because the absorption is very strong, light penetrates only into the outermost few hundred atoms. Furthermore, the interaction of this absorbed electromagnetic radiation with the highly conductive metal causes the rapid re-emission of the radiation without any change in the wavelength. This re-emission gives metals their glossy or lustrous appearance. In the majority of metals, the degree of absorption and re-emission does not vary across the visible spectrum, so these materials have a neutral colour – which we might term 'grey' or 'silver' or even 'white'. It is not a surprise that we use the colour term 'silver', as the metal is one of the best-known examples of such behaviour (Figure 2.15). Other commonly-encountered neutral-coloured metals include iron, zinc, tin and lead, all used since antiquity, and chromium and rhodium, both of which were used from the twentieth century to create an even more silvery appearance by plating one of them onto another metal.

For reasons connected to the selection rules governing allowed and disallowed transitions mentioned earlier, the efficiency of absorption and re-emission may decrease as the energy of the radiation increases, i.e., short-wavelength blue light is re-emitted less efficiently. As a result, metals displaying this property may appear more yellow, orange or red. The best-known examples are copper and gold (see Figure 2.15), and the many alloys containing these metals such as bronze (copper and tin); brass (copper and zinc); and red to pink alloys of gold with copper.

The highly reflective properties of metals are used not only to produce lustrous metallic objects, such as jewellery, vessels and mirrors, but also to juxtapose with painted surfaces, well-illustrated by the gold backgrounds in early Renaissance paintings and Orthodox icons, to provide a reflective base beneath transparent layers of coloured enamel or paint, or to add glitter when dispersed as powders in *maki-e* lacquered objects.

2.1.6 Band gap colours

In some elements and compounds, more than one discrete band of the type described above may be formed. This is illustrated in a theoretical example shown in Figure 2.16, where s and p orbitals are both involved in bonding. When the distance between atoms is great, the energy levels are discrete, but as the atoms approach each other, the electrons are shared in bonds and bands of energy levels

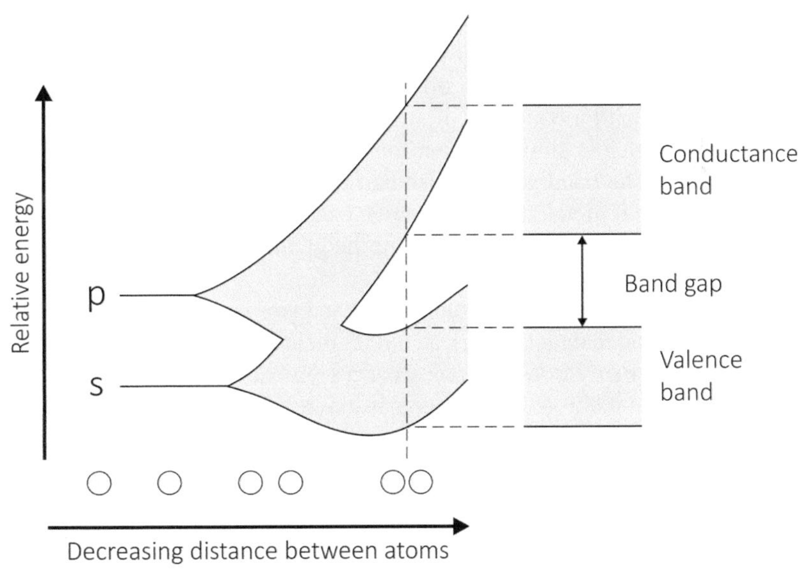

Figure 2.16 Schematic representation of the formation of separate valence and conductance bands, leading to the band gaps responsible for colour in some materials

corresponding to the two orbitals are created. While in metals these bands overlap to form a single continuum, in other elements and compounds there may be a gap between the lower (valence) and higher (conductance) energy bands. Such materials are termed semiconductors. If this energy gap – or **band gap** – corresponds to the absorption of light in the visible region, a coloured material may result.

The red colour of the synthetic pigment vermilion and its naturally occurring analogue cinnabar, both of which have been used since antiquity, derives from absorption of radiation to promote excitation of electrons across a band gap. In vermilion the band gap is 2.0 eV, which corresponds to a wavelength of *c*.620 nm. Radiation in the blue, green and yellow regions of the spectrum with wavelengths shorter than 620 nm, i.e., those that possess energy greater than the band gap, may be absorbed, while longer wavelengths, in the red region of the spectrum, will be reflected.

More recently introduced pigments whose colour also originates from band gap transitions are the cadmium yellows, oranges and reds. Pure cadmium sulfide (CdS) has a band gap of 2.6 eV, which equates to a wavelength of *c*.480 nm. Following the reasoning above for vermilion, radiation in the green (and particularly blue) region of the spectrum, with wavelengths less than 480 nm will be absorbed, while longer wavelengths (yellow, orange and red) will be reflected – giving CdS its characteristic yellow colour (Figure 2.17). A range of cadmium-based colours was introduced from the middle of the nineteenth century, becoming increasingly available to artists as their cost declined.

The shift in colour from yellow to red corresponds with increasing substitution of sulfur by selenium. Pure cadmium selenide (CdSe) has a band gap of 1.7 eV, which implies that all wavelengths in the visible will be absorbed, and it is indeed black. In compounds with an intermediate composition ($CdS_{(1-x)}Se_x$) the band gap decreases from 2.6 eV for pure yellow CdS (x = 0) towards 1.7 eV as x approaches 1 for pure CdSe. Many different hues are thus possible, for example, cadmium orange (with a band gap of around 2.3 eV), cadmium red, with a band gap of *c*.2.0eV, which is similar to that for vermilion (see Figure 2.17), and even cadmium brown. The band gap can also be tailored to produce a different colour by adding a different metal. The best-known examples are cadmium zinc sulfides ($Cd_xZn_{(1-x)}S$, with x in the range 0.75 to 1); the greater the proportion of zinc, the lighter the colour.

Other pigments that owe their colour to a band gap include the arsenic sulfides orpiment (As_2S_3) and realgar (AsS); lead oxides including red lead (Pb_3O_4) and massicot (β-PbO); and many iron-containing earth pigments, see Table 2.1. The observant will notice that many of the pigments listed in Table 2.1 are also mentioned in the section on ligand–metal charge transfer colours. This is because the band gap in these instances corresponds to that between the lower energy electronic levels for the ligands and those on the metal to which electrons are excited.

Many pure semiconductors have a band gap energy that corresponds to the absorption of wavelengths outside the visible region. For example, the band gap in a pure diamond is 5.4 eV, well outside the visible range, which is why pure diamonds are colourless. However, impurities in a diamond can give rise to colour.

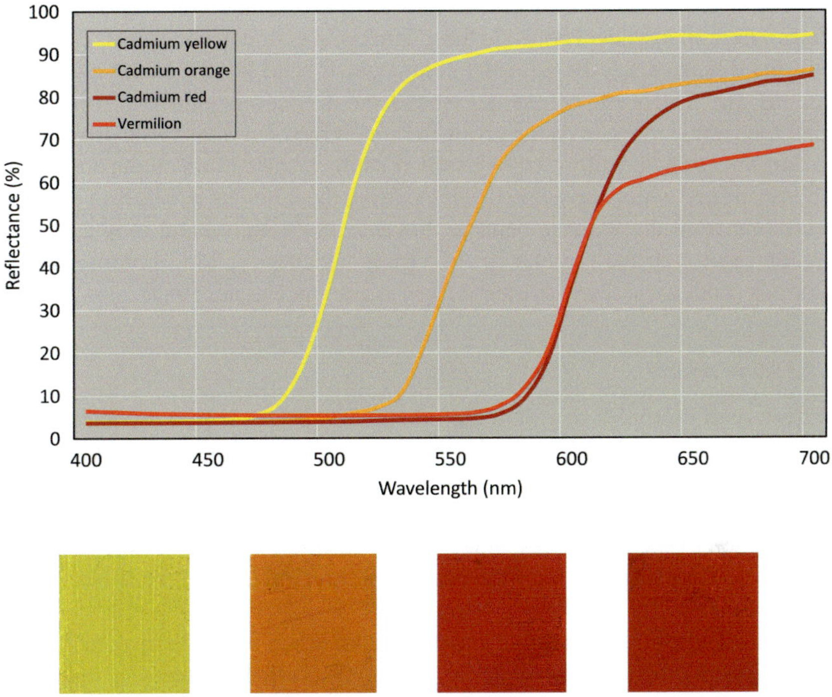

Figure 2.17 Reflectance spectra for four band gap colours, painted samples of which are shown below. Left to right: cadmium yellow; cadmium orange; cadmium red; and vermilion

Table 2.1 Some pigments that owe their colour to band gap transitions, with details of their colour and band gap energy

Pigment	*Compound*	*Colour*	*Band gap (eV)*
Vermilion/cinnabar	HgS	Red	2.0
Cadmium red	CdS/CdSe	Red	~2.0
Red lead	Pb_3O_4	Red	2.14
Hematite (red earth)	α-Fe_2O_3	Orange-red	~2.2
Cadmium orange	CdS/CdSe	Orange	~2.3
Realgar	AsS	Orange	2.5
Goethite (yellow earth)	α-Fe(OH)O	Yellow	~2.5
Cadmium yellow	CdS	Yellow	2.6
Orpiment	As_2S_3	Yellow	2.7
Massicot	β-PbO	Yellow	~2.8

If a few of the carbon atoms in a diamond are replaced by nitrogen atoms (which have one more electron than carbon and are thus termed electron donors), this creates a new energy level between the valence and conductance bands. Transitions from this donor energy band to the conductance band have a lower energy than those from the valence to conductance band and may correspond to the high energy end of the visible range, causing nitrogen-containing diamonds to absorb blue light and thus appear pale yellow. Conversely, if some of the carbon atoms are replaced by boron atoms, which have one fewer electron, the latter act as electron acceptors. Transitions from the valence band to this acceptor energy band may correspond to an absorption at the red end of the visible spectrum, giving boron-containing diamonds (both natural and synthetic) a blue colour (Nassau 1987: 16–17).

2.1.7 Colour centres

We have seen examples of where the substitution of atoms in a regular crystal or lattice structure gives rise to colour (in blue and yellow diamonds) and where an ion is held within gaps in the lattice – the case for ultramarine. The term 'colour centre' – or F-centre, from the German *Farbe*, which means colour – is sometimes applied to such instances, and even to transition metal ions held in a crystal lattice, for example, to the chromium ions that give rubies and emeralds their distinct colours.

Technically, however, the term should be used when there is a discontinuity in a lattice, typically the absence of an ion (sometimes called a 'hole'), or the replacement of a negative ion by an electron. As in the substitution of atoms in yellow or blue diamonds, these changes create an energy level intermediate between the valence and conductance bands opening the possibility of absorption in the visible region by otherwise colourless materials. Most examples of colour centres occur in minerals and gemstones, often created artificially by subjecting colourless materials to intense radiation (Nassau 1987: 18–20).

2.1.8 Absorption in organic molecules

Simple organic compounds, for example, the hydrocarbons, alcohols or ketones used to clean objects, are colourless. So, too, are the materials that we use as varnishes, coatings and adhesives; we regard this as an important property and are also careful to select those compounds that tend not to develop colour on ageing. In these molecules the energy levels are sufficiently far apart – particularly the HOMO and LUMO – that any electronic transitions correspond to the absorption of radiation outside the visible region. For a thorough description of the origins of colour in organic molecules, see Robert Christie's *Colour Chemistry* (Christie 2015).

The presence of double or triple bonds within organic molecules increases the chance that they will be coloured, particularly where two or more double bonds occur close to each other in the structure of the molecule. The underlying cause of this change in behaviour is that when these nearby double bonds in unsaturated compounds interact with each other (a process known as conjugation), the electrons that might be associated with individual bonds in the chains or rings become 'delocalised' across a larger portion of the molecule (see *Conjugation* box).

Conjugation

When two or more C=C double bonds are separated by one C–C single bond, the symmetry of their orbitals allows an interaction, termed conjugation, that results in the electrons in those bonds becoming delocalised over a larger part of the molecule, altering the molecular energy levels.
The two compounds below both have two double bonds in a six-carbon chain. However, in the molecule to the left (1,4 hexadiene) the two double bonds are separated by two single bonds and do not conjugate, whereas in 1,3 hexadiene, to the right, the double bonds are separated by only one single bond and are thus conjugated.

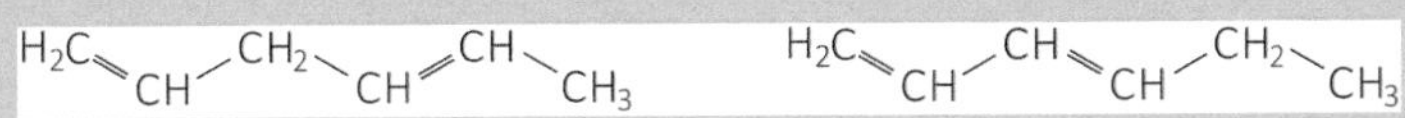

Other types of double bond, such as C=N, N=N or C=O can also take part in conjugation.

Conjugation changes the molecular energy levels and can create HOMO–LUMO energy gaps that correspond to the absorption of light in the visible range. Common examples of such compounds are those containing long chains of carbon atoms with many double bonds (e.g., the red biopigment lycopene found in tomatoes) or polycyclic molecules (those with a number of ring structures that include double bonds); some examples of the latter class are shown in Figures 2.18–2.20.

In these complex organic compounds it is often possible to identify a particular part of the molecule that is responsible for its colour and we refer to this as the chromophore (from the Greek *chromo* = colour and *phore* = bearing). In the dyestuff madder, the principal colorant is alizarin. The three-ring anthraquinone structure in alizarin is also the chromophore in the main coloured components in many other natural red colorants, including extracts from insects such as the kermes and cochineal beetles (see Figure 2.18). In a similar way, many yellow natural organic dyes are based on the same basic three-ring flavone/isoflavone unit (Figure 2.19) and the blue dye indigo and Tyrian purple (shellfish purple) share the same four-ring structure (Figure 2.20). For details of further organic colorant structures, see Cardon (2007).

Where the colour of the chromophore is modified by other parts of the molecule, the latter are termed auxochromes. In complex organic molecules, their presence alters the energy levels in the chromophore, subtly changing its absorption of light and hence colour. We might expect, therefore, to see slightly different colours for the three compounds in Figure 2.18 and will see in Chapter 5 that these changes

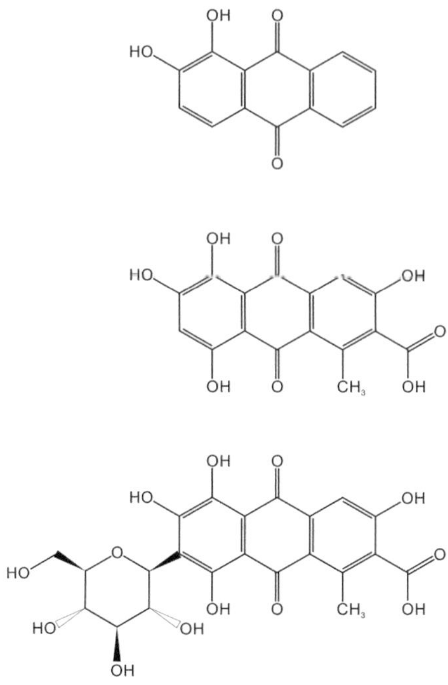

Figure 2.18 Molecular structures of three natural red colorants that owe their colour to the same three-ring anthraquinone unit

in the chemistry of the various anthraquinone-based colorants also affect their stability, particularly when exposed to light or ultraviolet radiation.

While the individual molecules that represent the principal coloured components in natural organic colorants derived from plant or insect matter are discussed above, in dyestuffs and pigments produced from these sources it is difficult to generalise about the overall colours (and indeed stabilities) of the resulting materials. This is because the compositions of the dyestuffs or pigments are strongly affected by many factors, particularly the method of their extraction from the raw materials and the manner in which they are bound to textiles when used as dyestuffs – i.e., the type of mordant used in the process – or the inert 'carrier' substrate onto which they are precipitated when they are used to create powder pigments, often termed lake pigments. For example, extensive studies of colorant production using the red anthraquinones in Figure 2.18 have illustrated that the method of dyestuff extraction, type of mordant, the temperature of the dyebath, the acidity or alkalinity of the dyebath and any additional materials added all influence the colour of the final pigment or yarn (Kirby, Spring and Higgitt 2005; Kirby, van Bommel and Verhecken 2014; Morales and Berrie 2015).

Modern synthetic colorants tend also to be based on conjugated organic molecules. The earliest of these, mauveine, was synthesised in the mid-nineteenth

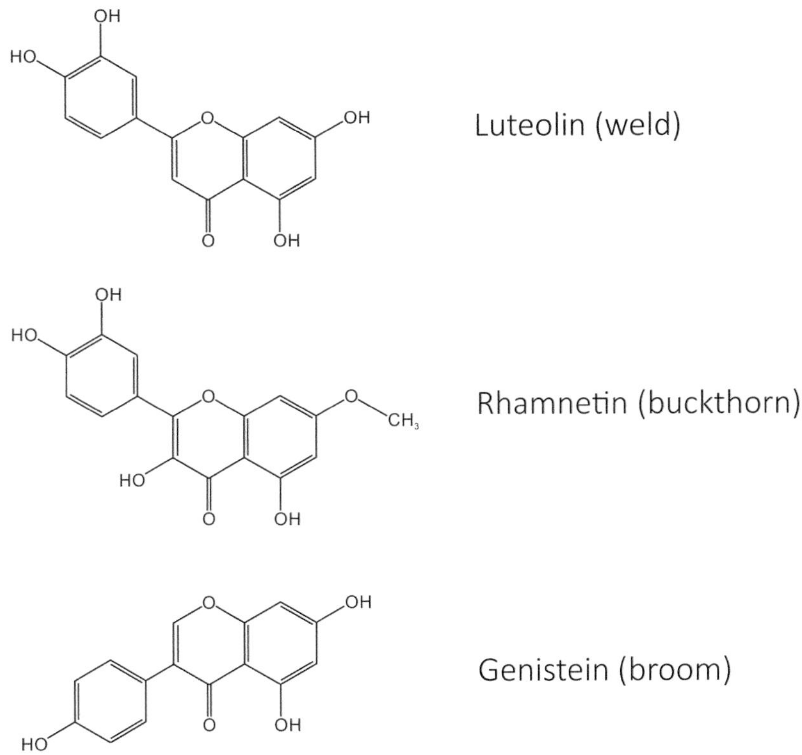

Figure 2.19 Molecular structures of three natural yellow colorants that owe their colour to the same three-ring flavone/isoflavone unit

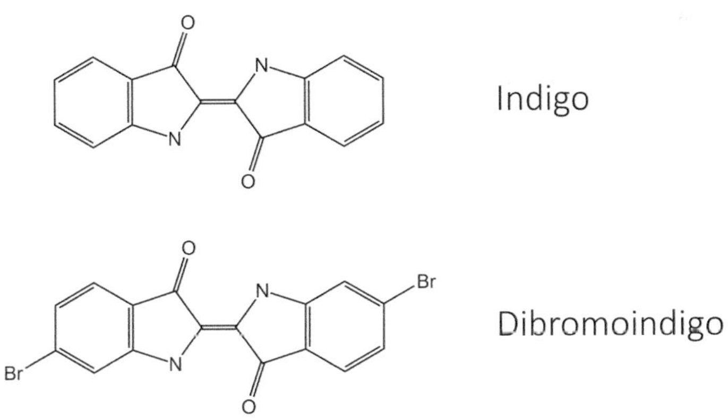

Figure 2.20 Molecular structure of the natural blue colorant indigo and closely analogous Tyrian purple (dibromoindigo)

century and there has since been an exponential increase in the number and range of organic colorants. Some of these colorants share a similar structure, with very small modifications to the molecule made to shift the colour or alter other properties. Here, the Colour Index provides a valuable resource (see *The Colour Index* box), as it assigns a unique reference number to each chemical compound, while a very good summary of the classes of modern organic colorants is provided by de Keijzer (de Keijzer 2014).

The Colour Index

The Colour Index™ specifies the precise nature of the colorant in a dye or pigment, defining its chemical composition and the way it is applied. Jointly maintained by the Society of Dyers and Colourists (SDC) and the American Association of Textile Chemists and Colourists (AATCC), the Index was first published in book form in 1924 and is now available online (https://colour-index.com).

In common with natural organic colorants, the chromophores in these synthetic materials tend to be based on unsaturated groups, such as C=C, C=O, N=N, C=N and NO_2. One of the most prevalent classes of synthetic colorants are those known as azo dyes, all of which contain the N=N group, generally surrounded by variously substituted aryl groups.[5] The azo pigments include para red (Colour Index reference PR1) toluidine red (PR3), lithol red (PR 49), permanent red (PR112), Hansa yellow (PY3) and tartrazine yellow (PY100), while diarylide yellows (e.g., PY12, 13 and 14) contain two N=N groups per molecule (Berrie and Lomax 1997).

Other synthetic organic colorants owe their colour to electron delocalisation across large polycyclic molecules. Some common examples are perinone red (PR194) and orange (PO43), dioxazine violet (PV23), red and orange pigments based on pyrrole and a series of red and violet pigments based on quinacridone.

Many modern blue and green pigments and dyes are based on phthalocyanine, the structure of which is shown in Figure 2.21. While phthalocyanine itself is blue, most phthalocyanine colorants contain a metal atom co-ordinated to the four central nitrogen atoms. Iron-based phthalocyanines were developed first (although their structure was not immediately recognised), but most modern pigments contain copper, including phthalocyanine blue (PB15) and its chlorinated analogue phthalocyanine green (PG7).

Modern organic dyes and pigments tend to be much cheaper to produce than those based on either metal salts or natural dyestuffs and generally show better

5 An aryl group is a group containing an unsaturated ring structure.

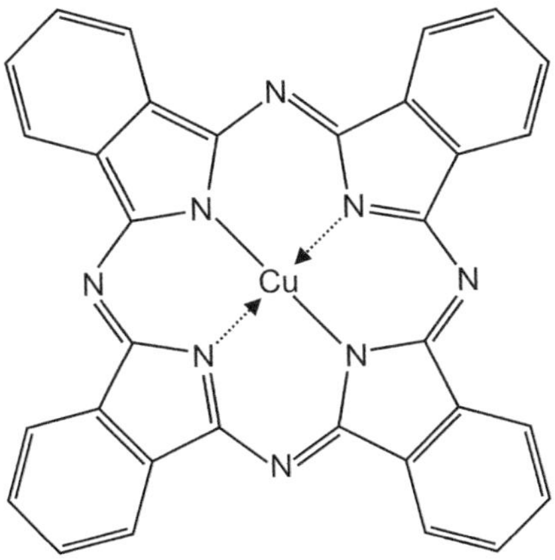

Figure 2.21 Molecular structure of the copper-containing phthalocyanine pigment (PB15)

stability. They are also generally less toxic both to human beings and to the environment than the more traditional metal-based pigments that they replace. For example, the cadmium-containing reds, oranges and yellows mentioned earlier are increasingly being replaced by mixtures of azo- and pyrrole-based colorants (e.g., arylide yellow, permanent red and pyrrole red), even if they are sometimes still described using colour names that hint at their original composition.

2.1.9 Biopigments

Most biopigments are also based on unsaturated organic compounds: indeed, we have already encountered the pigments and dyes produced by extracting and fixing red anthraquinones and yellow flavonoids from insects and plants. Other colorants may not be useful as pigments or dyes, but simply lend colour to the plants and animals that produce them. Generally, they owe their colour either to unsaturated long chain compounds, for example, the carotenoids, including the yellow–orange carotene, or to unsaturated polycyclic molecules, exemplified by melanins and anthocyanins.

Chlorophyll, which is responsible for the green colour in plants, contains a polycyclic aromatic structure in which magnesium is co-ordinated to four nitrogen atoms at its centre, making it conceptually, if not chemically, similar to phthalocyanine blue with its central co-ordinated copper.

One notable feature of many biopigments is that because they serve a function within the biochemistry of living organisms, they may react quickly to light and not be replenished after death, for example, we know that cut grass rapidly loses the green colour of chlorophyll.

2.1.10 Photoluminescence

We saw earlier – in the example of neon lights – that colour can be caused by the emission of radiation from an excited state, created in that instance when an electrical current is applied (a process termed electroluminescence, see *Other types of luminescence* box). Such excited states can also be created by the absorption of radiation, and this is the basis for the colour in many of the materials we have so far encountered in this chapter. In examining the origins of their colour, we have not yet looked at what happens to the excited states that are formed when radiation is absorbed and at the fate of the energy associated with that radiation.

Other types of luminescence

In addition to the concept of photoluminescence discussed here, colour can also be created by other processes that increase the energy of a molecule causing it to emit light as a result. These are:

- *Electroluminescence*: an electrical discharge through a gas can create light, for example, in neon lamps. When an electric current is passed through certain semiconductors, light is emitted, which is the principle of the LED lamp.
- *Radioluminescence*: emission of light following bombardment by ionising radiation. Found, for example, in older luminous paints containing a mixture of radium and copper-doped zinc sulfide used in watch and instrument faces that may form part of industrial or air and space collections
- *Chemiluminescence*: is produced by chemical reactions that create an excited electronic state which emits light as it returns to the ground state. The colour is usually relatively short-lived as the reactants are used up. Not generally a source of colour in cultural heritage collections, it is relevant to some studies of colorants interacting with substrates such as the paper to which they are applied.
- *Bioluminescence*: effectively chemiluminescence in living creatures but can be more persistent as the organism often has a method of regenerating the reactants. Not generally relevant in cultural heritage.

Light can also be produced by physical action (triboluminescence) or heat (thermoluminescence) but these are not sources of colour in collections, although the latter is used as a method of dating archaeological ceramics.

A certain amount of the radiation absorbed will be re-emitted at the same wavelength with no change in energy (Figure 2.22) and we saw earlier that rapid re-emission is responsible for the colour of metals. Another common fate for the energy associated with absorbed radiation is conversion to heat; on a macroscopic

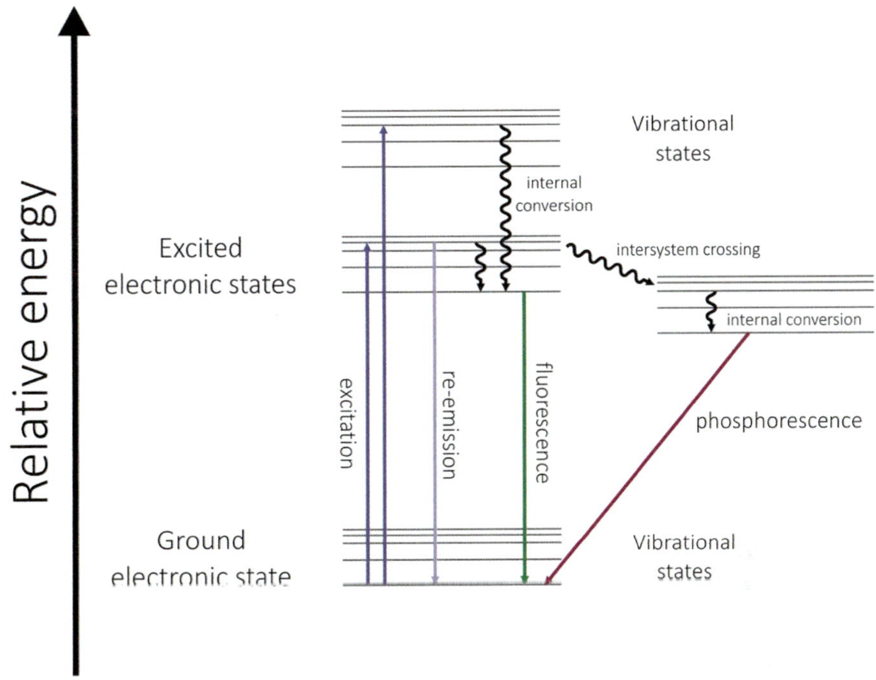

Figure 2.22 Schematic representation of the electronic excitations and emissions that give rise to fluorescence and phosphorescence

scale, we know that an object exposed to light can become warmer. Changes in vibrational or rotational energy state of the atom or molecule can also account for some absorbed energy.

Of interest in the context of colour, however, is the possibility that some of this energy is re-emitted as radiation at a different wavelength, a process known as photoluminescence. As the energy of the radiation emitted by a material will always be lower than that which causes this photoluminescence, the wavelength of the radiation emitted will be longer than that absorbed. So, for example, the absorption of ultraviolet radiation might cause luminescence in the lower-energy, longer-wavelength visible region. As well as wavelength, a second property used to characterise photoluminescence is its lifetime, the average time between excitation and re-emission, which typically varies from femtoseconds (10^{-15} seconds) to milliseconds, with some occasional examples of even longer lifetimes.

There are two common photoluminescence processes that contribute to the colour of certain objects: fluorescence and phosphorescence. Phenomenologically, the former is characterised by much shorter lifetimes, typically in the order of nanoseconds, while the latter processes have longer lifetimes, with emission often persisting once the source of excitation has been removed. At a molecular level

fluorescence is the result of re-emission from the state to which the molecule was first excited, after the dissipation of some energy through non-radiative processes, such as vibrational relaxation (see Figure 2.22).

Phosphorescence occurs by re-emission from a third energy state; the emission is much slower because once the molecule reaches this state – by a process termed intersystem crossing – emission is forbidden by the quantum mechanical rules that govern energy level transitions (see Figure 2.22).

The terms fluorescent and phosphorescent are not always applied with great accuracy, for example, fluorescent lamps rely principally on phosphorescence to convert ultraviolet radiation emitted by the electroluminescence of mercury vapour to visible light. Regardless of what we term the process, by varying the phosphors coating the inside of the glass envelope, fluorescent lamps can be tuned to emit a range of colours of light (see Figure 2.3).

In addition to colours created wholly by photoluminescence, such as those in some light sculptures, there are examples of materials whose colour is altered by luminescence, sometimes called fluorescent colours. In these cases, the reflected radiation is augmented by emission at certain wavelengths in the visible region caused by the absorption of either shorter wavelength visible light or ultraviolet radiation. Historical examples of such materials include certain anthraquinone dyestuffs and the pigment Indian yellow (also known as *puri*) (Tamburini et al. 2018). There exists a range of modern pigments formulated to take advantage of fluorescence generated by the absorption of ultraviolet radiation to give intense, striking colours (Figure 2.23), sometimes inaccurately described today as Day-Glo colours from the name of a successful product line in the 1960s. Although the materials illustrated here show strong fluorescence, in other cases, the emission is so weak that it contributes little to the perceived colour, but can be useful in identifying the material when the fluorescence is viewed or measured in isolation, see Chapter 3.

Fluorescent colours are not limited to pigments. Many mineral specimens also show fluorescence and, as mentioned earlier, glasses containing uranium show a pale yellow colour due to f–f absorptions, but can also demonstrate a green luminescence (see Figure 2.11).

A final example of fluorescence in works of art is the use of so-called **optical brighteners** in modern papers. These materials are designed to counter the tendency of paper to have, or acquire on ageing, a slightly yellowish hue. They function by absorbing radiation in the ultraviolet region of the spectrum and re-emitting at the blue end of the visible spectrum. This 'additional' blue fluorescence can serve to balance any yellowish cast in the paper substrate.

Fluorescent colours – and optical brighteners – present an interesting conundrum when displaying works that contain them, as their appearance will depend on the type of light with which they are illuminated. In museums, where ultraviolet radiation is often specifically excluded, the fluorescence will be absent or much reduced and the appearance may be different from that intended by the artist when creating the work under conditions in which ultraviolet was present, for example, in unfiltered daylight. As we will see in Chapter 5, a further complication is that

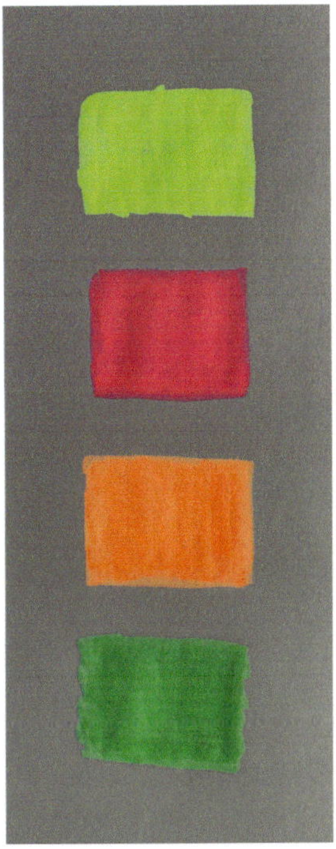

Figure 2.23 Samples of modern yellow, pink, orange and green fluorescent colorants. Photographed in daylight to ensure ultraviolet excitation produces visible fluorescence

many fluorescent colours are also rapidly degraded by visible light and ultraviolet radiation.

2.2 Physical causes of colour

2.2.1 Structural colours

Some colours result from the way in which white light interacts with the surface of an object, specifically the morphology of that surface, which is why such instances are often termed structural colours. To appreciate how these colours are produced we first need to understand the process of **interference**.

In Chapter 1 it was stated that light and other types of electromagnetic radiation can be considered to propagate as waves, to which we can assign a wavelength that is characteristic of each type of radiation (or in the visible region, to each colour of light). We can represent such a wave diagrammatically (Figure 2.24).

The wave is characterised by a series of crests and troughs, with the distance between successive crests (or troughs) equal to the wavelength of that radiation.

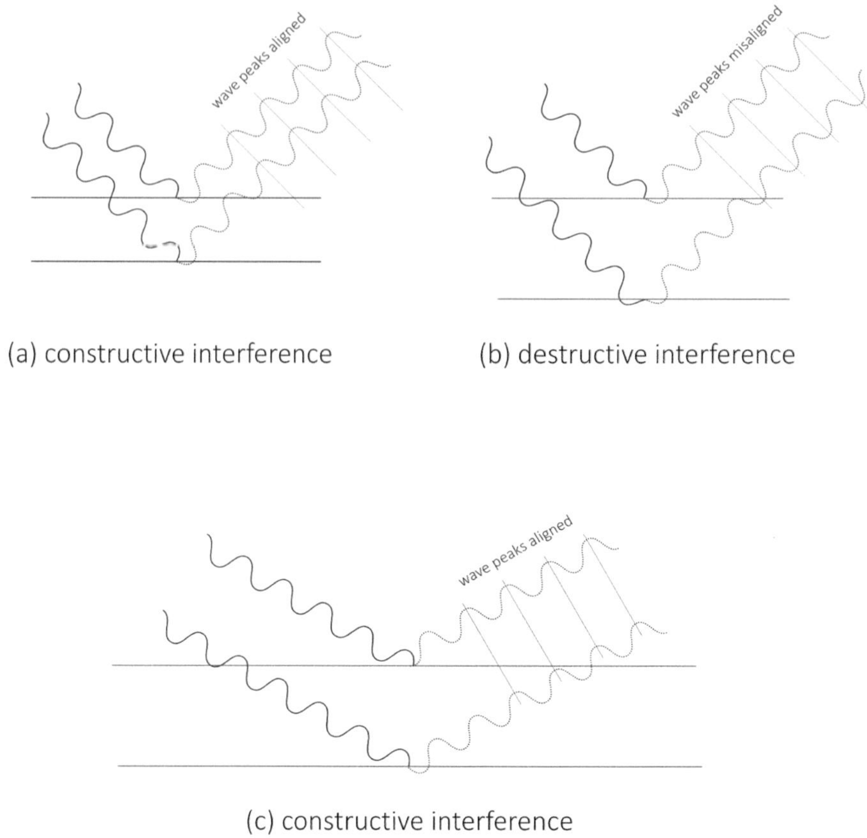

Figure 2.24 Schematic representation of the conditions leading to constructive and destructive interference

As we saw in Chapter 1, all electromagnetic waves propagate at the same speed (termed the speed of light). Let us consider two waves of the same wavelength reflected at two different surfaces. Under some circumstances, the distance travelled by the two waves will differ by a whole number of wavelengths, so that the peaks of the two reflected waves will coincide in the reflected beam that reaches the eye (Figure 2.24a). The waves are said to be 'in phase' and this wavelength of light will be reinforced, a phenomenon termed constructive interference.

In contrast, if the difference in distance travelled by the two waves corresponds to a half of the wavelength (or 1½, 2½, etc. wavelengths) the peaks of one wave will coincide with the troughs in the other and the waves will be out of phase (Figure 2.24b). This will lead to a cancelling out of this wavelength through destructive interference. As a result, this wavelength will be absent from the reflected beam and we will observe its complementary colour.

Classic examples of the phenomenon are the interference colours observed from soap bubbles or a thin film of oil on water. Interference arises from the difference in the distance travelled by light (termed the path length) that reflects from the inner or outer surface of the bubble or oil film. We know that as we alter our viewing angle the colours we observe change. This is because the path length changes with angle – light entering and leaving the film at shallow angles travels further – so different wavelengths will constructively or destructively interfere as the angle changes. We can see this diagrammatically in Figure 2.24c; here the wavelength and the spacing between surfaces are the same as in Figure 2.24b, but the change in angle creates a longer path length through the film, which in this case causes the wave peaks in the reflected beam to align, giving rise to constructive interference.

In the cultural heritage sector, the most common structural colours are associated with natural history specimens, for example, in some birds' feathers, insects' wings and molluscs. In these cases, the interference colours arise either from reflection at the surfaces of extremely thin structures in the wings and elsewhere (analogous to the colours seen in soap or oil films), or from reflection or scattering from a structured surface. In the latter case, the wavelengths of light that experience destructive interference (and hence are absent from the light seen by the viewer) or which are reinforced by constructive interference, will depend on the distance between these structural elements. In the striking example of Morpho blue butterflies, modified scales on the upper wing surfaces show a periodicity of around 200 nm, creating constructive interference in the blue region of the visible spectrum. Beetle wing cases (elytra) that produce a green interference colour were widely used in Basohli miniature paintings of the late seventeenth century to represent emeralds (Eisen 2023), and have also been incorporated into Indian textiles (Figure 2.25).

It is worth noting, however, that the colours in feathers and insect wings can arise from both structural coloration and the presence of pigments. Generally, yellow, red, and brown colours are due to some of the biopigments (e.g., carotenoids and melanins) mentioned above, while structural effects produce green and blue (Riedler et al. 2014: Table 1).

The transparent, lamellar structures of nacre (mother of pearl) and the mineral mica give rise to interference (iridescent) colours. Nacre is an example of iridescence in natural history specimens, but the material is also popular as an inlay for furniture and lacquerware, and, when ground, has been used to provide an iridescent pigment or to make papers that provide the glimmering background in certain prints. Mica and similar minerals have also been used to add iridescent colour to objects and a range of modern pigments contain mica particles that add iridescence and transparency.

A final example of interference colours is offered by one of the first methods of colour photography, developed by Gabriel Lippmann in the late nineteenth century and for which he won the Nobel Prize in Physics in 1908. The process relied on placing a temporary (mercury) mirror behind a silver halide-treated glass photographic plate. The interference patterns created by interaction of the direct and

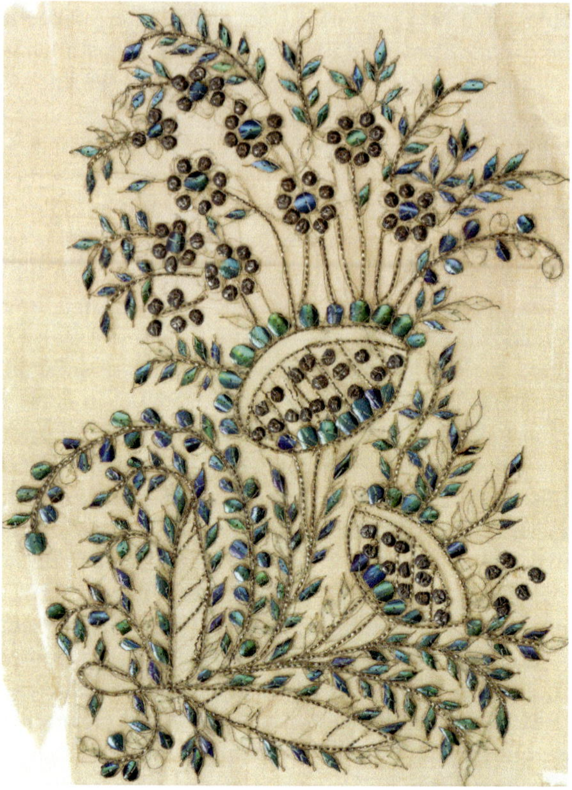

Figure 2.25 Blue and green interference colours created by beetle wing cases (elytra) incorporated into an embroidered cotton textile fragment. [Source: India, mid-nineteenth century. 24.2 x 17.3 cm. Cooper Hewitt, Smithsonian Design Museum, 1931-43-20. Gift of Sarah Cooper Hewitt.]

reflected wavefronts created, after processing, silver particles in a lamellar structure, with the spacing between the lamellae corresponding to the wavelengths of the light incident on that area. When viewed, interference at these surfaces creates colour, although there are some doubts about the extent to which the colour seen by the viewer represents that of the original scene (Baechler et al. 2021).

2.2.2 Colour due to scattering

We saw in Chapter 1 how scattering can affect the appearance of colours, particularly the transparency or opacity of materials, but here we will explore how it can give rise to colour. Scattering arises as radiation moves from one medium to another – for example, at the interface between the air and the surface of an object, or between a paint medium and pigment particles held within. Some scattering results in a change in energy between the incident and scattered radiation. This is termed inelastic scattering, and one of the most frequently encountered instances is that of Raman scattering, the principle behind which is the basis of Raman spectrometry for the identification of materials.

Colours derive, however, from elastic scattering of light, where there is no change in the energy or wavelength of the scattered radiation. The colour arises because different wavelengths of light are scattered to different extents so that – rather as in the case of interference colours – the ratio of light of different wavelengths is changed in the scattered beam that reaches the viewer.

Formally, the relationship between scattering (S) and wavelength (λ) is $S = k/\lambda^4$, where k is a constant, but in straightforward terms, the shorter the wavelength, the greater the scattering. Within the visible spectrum, blue light will be scattered more than red, a phenomenon that accounts for the sky appearing blue and sunsets red.

Several types of elastic scattering are encountered in the literature, including Rayleigh, Tyndall and Mie scattering. The name Rayleigh tends to be applied to scattering by very small particles, much smaller than the wavelength of radiation concerned (and indeed to scattering where no particles are present). In Tyndall scattering, the particles are of approximately the same size as the wavelengths of light involved. Mie scattering is a particular case of scattering for spherical particles of any size. In each case the term 'particle' can be misleading as scattering can occur from other features that give rise to changes in refractive index, such as small bubbles or pockets of air.

Colloids are materials in which microscopic particles of one substance are dispersed throughout a second substance, and there are several examples of coloured colloidal materials, as we will see below. Many sources attribute colloidal colours to the selective scattering of light of different wavelengths by these very small particles, as described in the previous section.

However, extensive research into nanoparticles (particles with a dimension of the order of nanometres) in the last few decades, mainly focused on the medical and pharmaceutical uses of gold nanoparticles, indicates that their colour does not arise from simple elastic scattering. Rather, the colour is due to the phenomenon of surface plasmon resonance, a complex interaction between electrons in these metal nanoparticles and incident radiation. This interaction relies on the size and shape of the metal nanoparticles and on the medium in which they are dispersed (Jana, Ganguly and Pal 2016). For example, as the diameter of gold nanoparticles dispersed in water increases from 10 to 90 nm, the observed colour changes from red to violet.

Small (10–30 nm) particles of gold are responsible for one of the most notable examples of colloidal colour encountered in the cultural heritage sector, a red glass sometimes termed ruby glass (Figure 2.26). This glass was developed in antiquity although, in common with some other materials, its production waned towards the end of the first millennium CE and revived when new methods of synthesis were developed in the seventeenth century (Louis 2017), and sometimes later called cranberry glass. Not all 'ruby' or 'cranberry' glass is, however, based on colloidal

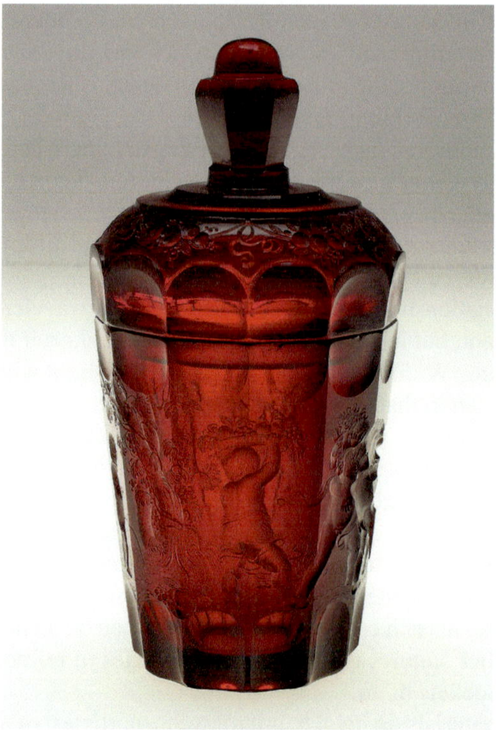

Figure 2.26 Ruby glass beaker with cover [Source: Gottfried von Spiller, 1685–1700 18.6 x 9.4 cm. Art Institute of Chicago, 1927.1208a-b. Gift of Julius and Augusta Rosenwald.]

gold; copper was the predominant colorant in the flashed red glass[6] of mediaeval stained-glass windows and modern glasses use copper and a number of cheaper colorants, including selenium.

At around the same time that a new synthesis of colloidal gold glass was developed, a new red-violet pigment was produced (reportedly by Andreas Cassius in 1676) by precipitating gold particles onto a tin(IV) oxide substrate. This pigment, known as purple of Cassius, was expensive, but was used in miniature painting and in ceramic glazes until it was replaced in the mid-nineteenth century by the cheaper cobalt-based pigments described earlier (CAMEO n.d.).

Nanoparticles of silver also demonstrate surface plasmon resonance, giving rise to a strong yellow coloration that again varies with particle size and shape. Prominent examples of yellow colloidal silver occur in 'silver stained' glass

6 Glass produced by flashing (layering) a thin sheet of intensely coloured red glass onto clear glass to produce a sufficiently transparent red colour.

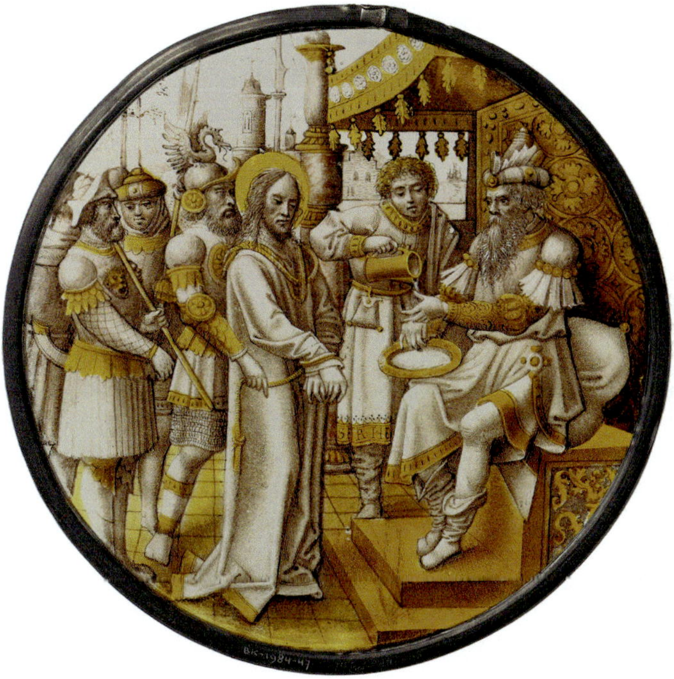

Figure 2.27 Southern Netherlandish glass panel coloured with silver stain, depicting *Christ before Pilate* [Source: Aert Ortkens (possibly): *c.* 1510–1530. c.22.5 cm Ø. Rijksmuseum, BK-1984-47.]

panels from the early fourteenth century onwards. A paint containing a soluble silver compound was applied to decorate clear glass. On firing, colloidal silver particles are created within the glass in the treated areas yielding a yellow colour (Figure 2.27).

2.3 Particle size

We have seen that the particle size in a colloid is critical in determining its colour. In pigments that owe their colour to their chemistry, the particle size is not the origin of the colour but may affect its strength. When mineral pigments are ground before mixing with a binding medium, it is often noted that their colour becomes lighter and less strong the more they are ground (Figure 2.28). As the particle size decreases, the ratio of surface area to volume increases so that diffuse reflectance at the air–pigment interfaces causes a greater proportion of white light to reach the observer. When pigment particles are mixed with a binder, particle size continues to play a role, as scatter at the interface between pigment and medium will be greater per unit volume for smaller particles, tending to produce lighter shades.

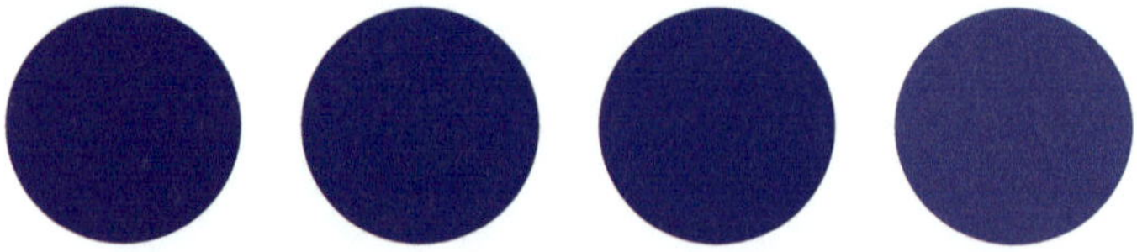

Figure 2.28 Samples of azurite pigment ground to produce increasingly finer particles from left to right, with a consequent increase in lightness

2.4 Combining colours

In looking at the origins of colour we have so far considered individual colorants. While – as we have seen – these provide a wide range of possible colours, makers or artists will often mix colours, perhaps to match more exactly the appearance of an object or scene they wish to depict. Section 1.8 introduced both additive and subtractive mixing, explaining how new hues arise from the combination of primaries. These mixed hues are frequently further modified by the addition of white or black pigments to tailor their lightness or darkness. Where pigments and dyes are mixed, or where one dyestuff is applied on top of another (e.g., in overdyed textiles), the final appearance will result from subtractive mixing. Less frequently we will need to consider the additive mixing of colours, for instance, when assessing the colour of light- or video-based works.

In objects where the final colour results from the more than one layer of colorant – or the application of a base or priming layer beneath the final paint film, the overall colour appearance may be influenced by the underlying colour, depending on the transparency and thickness of the outer layer (its hiding power); in Figure 1.9, we can see that the observed colour in the transparent paint film is altered in areas where there is underlying black text.

Although a detailed description is beyond this introduction to colour, it is possible to model the appearance of such layered colours mathematically, using the Kubelka–Munk (K–M) theory (Kubelka and Munk 1931). Although first developed to determine the effect of a coating on the colour of an underlying layer by considering the remission (reflectivity or back-scatter) and absorbance of the coating, it is also possible to use K–M theory to model the effect of mixing pigments (Dichter 2023).

3 The measurement of colour

How we measure colours

The spectra for coloured compounds contain a great deal of information about the reflection and absorption properties of objects and, as these are related to the molecular structure of materials, they can be useful for material identification. We will look at how spectrometry (or spectrophotometry[1]) can be used to measure colours in Section 3.2. Historically, however, it has only been possible to produce spectra for under 200 years and for much of that period the spectrometers used for this purpose were complex, expensive and unwieldy instruments. In addition, even for the experienced user, spectra do not give particularly straightforward information about how a colour appears to us.

3.1 Colour matching

Before the advent of instrumental methods, the classification of colours often relied on comparison with colour standards or samples in colour atlases. Colour standards, which have been used in dyeing, textile manufacture and other industries where matching a shade is important, have been employed for several centuries, but they are not easily communicable, as colour information can only be conveyed by passing the standard from one user to another, or by making exact physical copies. Early printed compendia of colours, such as *Werner's Nomenclature of Colours* (Syme 1821), and shade cards of pigments or dyes relied on hand-colouring and it was not until the advent of good-quality colour printing in the nineteenth century that it became possible to produce multiple mechanical copies of colour samples. Even then, relatively small numbers were produced, and the process was complex, pushing printing technology to its limits, as, for example, in the publication of Michel Eugène Chevreul's chromatic circles (Chevreul 1864).

In the twentieth century, colour atlases became more common and several new colour order systems were developed that attempted to semi-quantify colour,

1 The terms spectrometry and spectrometer, applicable to any wavelength range, have been used rather than spectrophotometry and spectrophotometer, which apply to the use of visible light only, except where the latter are more commonly applied – as for microspectrophotometry – or where the term spectroscopy is more usual for a particular technique.

DOI: 10.4324/9781003397182-4

assigning an alphanumerical classification to each shade. For example, in the mid-twentieth century, a system developed by Wilhelm Ostwald enjoyed great popularity, not least because it was espoused by the pigment manufacturers Winsor and Newton (Ostwald 1916; Scott Taylor 1935). The Natural Color System (NCS), developed in the 1970s (Hård, Sivik and Tonnquist 1996a; 1996b), continues to be used in the paint industry to define colour, but the colour atlas most frequently encountered in the cultural heritage sector is that developed by Albert Henry Munsell in the early twentieth century (Munsell 1905).

In common with the earlier colour order systems described in Chapter 1 and Chevreul's chromatic circles, both the NCS and Munsell systems are based on a circular arrangement of different hues, each defined as a primary colour or mixture of two primaries (of the four in the NCS system or five in the Munsell). While Chevreul added increasing amounts of black to produce darker shades, the Munsell and NCS systems have a full range of lightness values from black to white (Figure 3.1). The innovative aspect introduced by Munsell (later used in the NCS system) was the concept of **chroma**, or saturation, which defines the strength of each colour: low chroma colours are close to the neutral grey axis in Figure 3.1, while high chroma colours are further away. Accordingly, colours in the Munsell system are assigned three 'co-ordinates', a hue, a value (the term used to express lightness) and a chroma (Figure 3.1).

In theory, this allows a unique description of a colour to be communicated without the need for a physical atlas of colour swatches but, although many modern spectrometers and colorimeters – described in the sections that follow – can provide hue, value and chroma in Munsell co-ordinates, the system is still most commonly used when it is advantageous to make a direct comparison of colour with a 'chip' from the *Munsell Book of Color*.

There are two common situations in which this might be the case. First, when institutions, not all of which have access to spectrometry or colorimetry, wish to compare colour data (although it should be borne in mind that the cost of a full set of Munsell chips is comparable to an inexpensive colorimeter): a good example is a study of the colour of threads preserved in Andean workbaskets held by four museums (Price et al. 2015).

Second, in harsh environments, including archaeological sites, where power may be limited and electronic instruments might be damaged or operate poorly. For example, the *Munsell Soil Color Book* and the *Munsell Bead Color Book* (both subsets of colours from the comprehensive atlas) are sometimes used in field archaeology to classify the colour of excavated finds (Ferguson 2014).

If colour chips are used to classify colours, it is important to have good conditions for viewing, as we are using our visual system as the 'instrument': those who have colour vision anomalies would need to assess the extent to which this affects their ability to make good matches. While it is often recommended that comparisons be made under daylight, we know that the colour of natural light can vary according to season, time of day and atmospheric conditions, so it is far from a constant reference. As we saw in Chapter 1, colour adaptation helps us to adjust to different white light sources, so the type of light is probably not the most critical factor. Diffuse

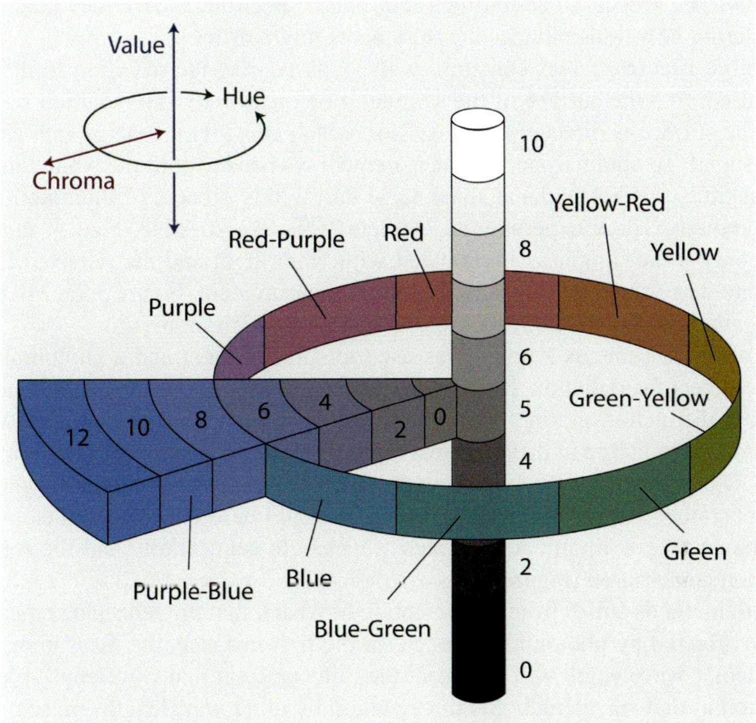

Figure 3.1 The Munsell colour system, showing the properties of hue, value, and chroma [Source: Jacob Rus, available at: https://commons.wikimedia.org/wiki/File:Munsell-system.svg]

rather than strongly directional light will help to reduce the interference of gloss in the matching process, but the most important influence on visual comparison is light level. Research from the 1960s indicated that good colour matches can be achieved once the light level exceeds around 10 lux, although if the level is higher, the process is more comfortable (Verriest, Buyssens and Vanderdonck 1963). The figure of 10 lux is for viewers in their twenties and subsequent research has shown that for older people the amount of light needed to complete the same task can be much higher – ten times more for those in their fifties and even greater for those in their sixties or seventies (Knoblauch et al. 1987).

3.2 Spectrometry

Turning to instrumental and quantitative methods of colour measurement, we might first consider the use of spectrometry. This measures the reflectance, absorbance or transmittance – depending on the type of material and why we require measurements – of an object at regular wavelength intervals throughout the visible region. The spectra shown in, for instance, Figures 1.5 and 2.17 were made in this way. The process of acquiring a reflectance spectrum is described here before considering how transmittance measurements might differ.

Visible spectrometers contain a white light source, the radiation from which is focused onto the surface of the sample to be measured. The radiation reflected from the surface is directed onto a sensor, which converts its energy into an electrical signal. To obtain a spectral measurement we need to split the white light into its constituent wavelengths at some stage during this process of illumination and measurement. Three experimental configurations are possible. First – and most commonly – the sample is illuminated with white light and the reflected light is split into its component wavelengths before measurement, Figure 3.2a. Microfade testing, to be discussed in Section 6.6, uses this configuration.

In older instruments a prism was used to split the light and a photomultiplier tube to detect the radiation, but modern instruments replace the prism with a more compact diffraction grating and tend to use a charge-coupled device (CCD) to measure the radiation at different wavelengths simultaneously (the position along the CCD corresponds to the wavelength: Figure 3.2a). In the second configuration, which is rather less common, the white light is split before it reaches the object so that the surface is illuminated at each wavelength sequentially and the reflected radiation is measured (Figure 3.2b).

Both methods suffer from the potential drawback that the reflectance measured can be affected by photoluminescence. In the first instance, the measurement for a particular wavelength will represent the reflectance at that wavelength plus any emission at that wavelength due to excitation by other wavelengths present in the white light used to illuminate the sample. In the second configuration, the measurement is the sum of the reflectance at the illuminating wavelength plus emission (at any wavelength within the measurement range of the detector) caused by radiation at the illuminating wavelength.

A third configuration avoids the effects of luminescence, by selecting a particular wavelength of radiation to illuminate the object AND selecting the same wavelength between the object and detector, so that luminescence (which will occur at longer wavelengths) is excluded (Figure 3.2c).

The added complication of selecting the wavelength before and after reflection at the object surface means that this third type of spectrometer is more expensive and complex, and so less frequently used for simple colour measurements. It can, however, be used to make precise measurements of photoluminescence, as the emission across the spectrum can be measured for each wavelength of incident, exciting radiation (Figure 3.3). The excitation–emission spectra produced by this technique, termed fluorimetry, can be helpful in material identification.

The same three experimental configurations can be employed to make transmittance or **absorbance** measurements (see *Reporting reflectance, transmittance and absorbance* box). Again, that used most frequently involves passing white light through the object before splitting the transmitted radiation into its component wavelengths and measuring the proportion transmitted at each wavelength.

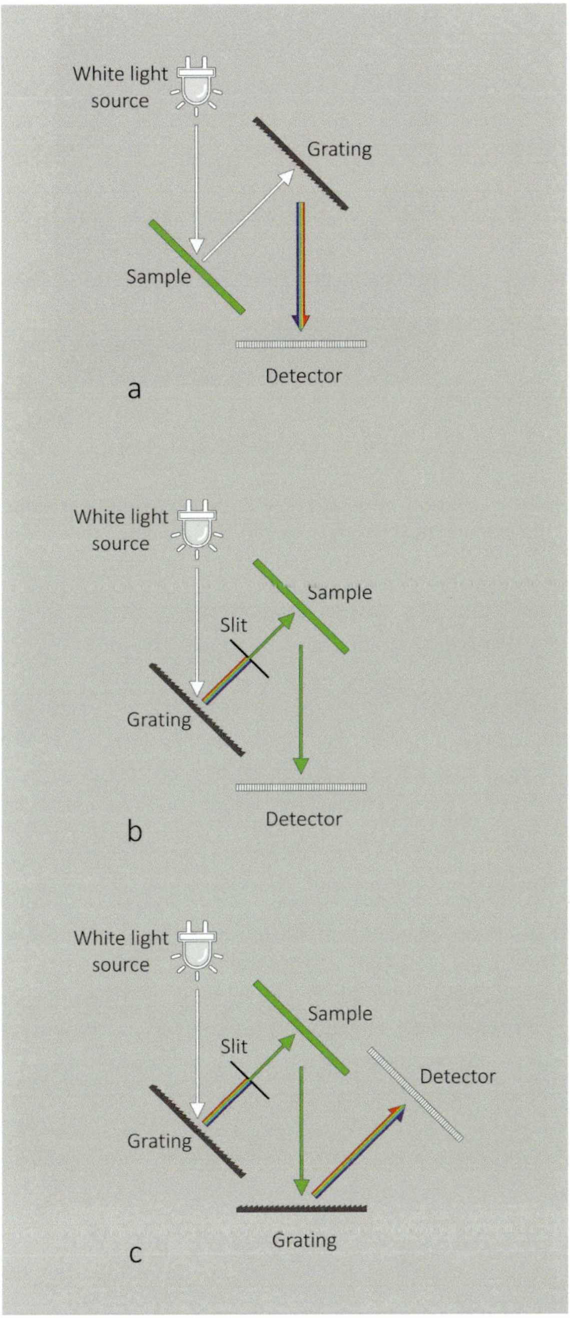

Figure 3.2 Schematic representations of the three types of reflectance spectrometer described in the text

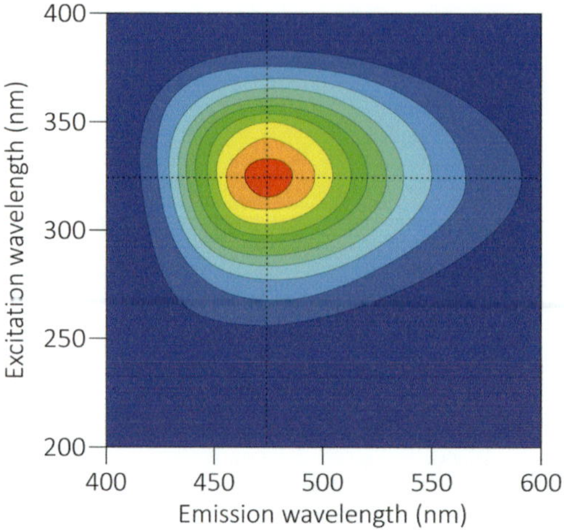

Figure 3.3 A typical excitation–emission spectrum showing a maximum emission at *c*.475 nm excited most strongly by radiation at *c*.325 nm

Reporting reflectance, transmittance and absorbance

Reflectance spectra present the reflectance at each wavelength as a proportion of that of a perfect white. This is usually reported either as a percentage – with 100 per cent representing a perfect white and 0 per cent the perfect black determined by the dark calibration – or on a scale from zero (black) to one (white).

In the same manner transmittance is generally reported either as a percentage – with 100 per cent representing the complete transmittance of light (through air) and 0 per cent total blocking of light, determined from the dark calibration – or on a scale from zero to one.

Absorbance values (A) are often reported by instruments automatically, but can also be calculated – if required – from the percentage transmittance data (T, the Greek capital letter tau), using the relationship:

$$A = -\log_{10}\left(\frac{T}{100}\right)$$

This scale of absorbance gives, for example, values of 0 for 100 per cent transmittance (no absorbance), 1 for 10 per cent transmittance and 2 for 1 per cent transmittance.

3.3 Spectrometers: Making quantitative measurements

If we wish to produce quantitative and reproducible spectra that can be used for certain purposes, such as colour difference measurement or to make tentative identifications of materials, we will need to control a number of factors in the measurement process.

3.3.1 Wavelength calibration

Before making any measurements, we must be certain that the wavelength scale of the spectrometer is correct. This is generally achieved using monochromatic laser radiation; a red helium-neon laser with an emission at 632.8 nm or a green laser at 532 nm are commonly employed. Some instruments allow the wavelength scale to be recalibrated based on this step, but even when this is not possible, determining the measured wavelength of the laser radiation gives an indication of the accuracy of the wavelength scale in subsequent measurements. This is more important if we wish to identify materials from their characteristic absorptions in the visible or surrounding spectral regions.

3.3.2 Wavelength resolution

When selecting a spectrometer, we may also need to decide whether it has sufficient wavelength resolution to address the questions we wish to answer, a topic that is covered again in Section 6.1. Many of the materials we encounter in cultural heritage objects have rather 'smooth' spectra, i.e., their reflectance or transmittance does not change quickly across the visible wavelength range and there are few peaks or troughs in their spectra. The spectra in Figure 1.5 are good examples. In such cases it might be sufficient to record the reflectance or transmittance at 10 nm intervals as is common in hand-held spectrometers. An interval of 5 or 10 nm is typically used when converting spectra into colorimetric data.

There are, however, certain cases where we might want to see the structure of the spectrum in more detail, particularly if we are attempting to use the information to aid material identification. In these instances, a resolution of 1 nm or less is quite common.

3.3.3 Reflectance or transmittance calibration

Next, we need a means of ensuring that the spectra are independent of the light source used during measurement, which will vary from instrument to instrument. In some instances, this may be achieved by using a so-called dual beam instrument. Here, light from the source is split into two beams, one of which is directed to the sample and the other to a reference standard. For reflectance measurements, the second beam is generally directed at a white reference material with a known spectrum. The reflectance of the sample is then compared to that of this white material at each wavelength. For transmittance and absorbance measurements on solids (e.g., glass or plastic objects), the second beam simply passes through the air,

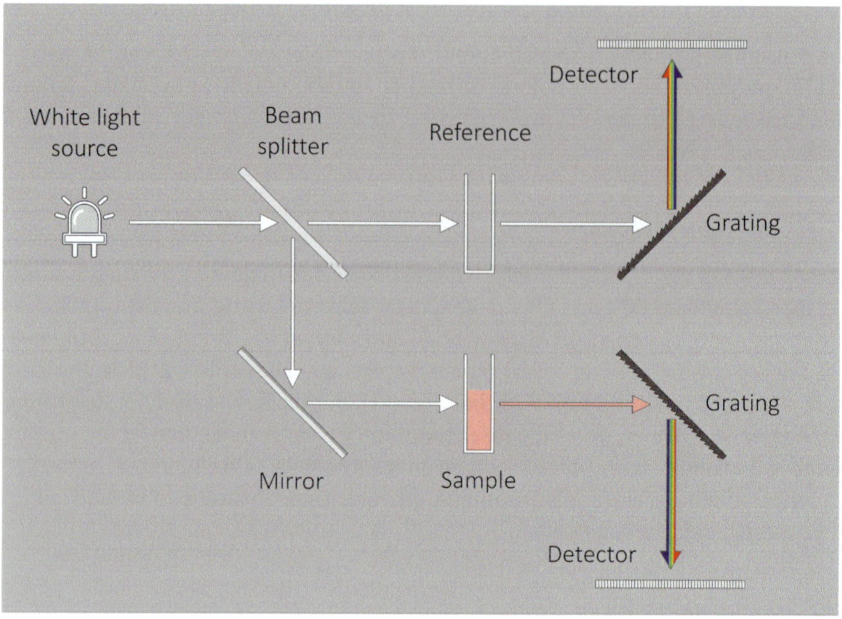

Figure 3.4 Schematic representation of the configuration of a dual beam transmittance spectrometer

before a comparison is made between the signal at each wavelength for the sample and reference beams. If the transmittance or absorbance of liquids or gases is to be measured, then the reference beam will pass through an empty container identical to that in which the sample is held (Figure 3.4).

Dual beam instruments are less common now than before the advent of modern instruments that use microprocessors to store and compare spectral data; they are generally found only in laboratories making very precise measurements. Most current spectrometers are single beam instruments, which make the comparison between the reflectance, transmittance or absorbance of the sample and reference materials consecutively rather than simultaneously. The usual procedure is to first record the spectrum of the reference material (white reflectance standard, empty container, etc.) and to store these data digitally in the spectrometer's microprocessor.

The white reflectance standards currently used in spectrometry and colorimetry are typically a piece of Spectralon, a matt white material made by compressing powdered PTFE (polytetrafluoroethene: Teflon). Traditionally a block of barium sulfate was used but has been replaced by these more durable polymeric standards. Reflectance standards are usually supplied with calibration data, produced by the manufacturer by comparing them with the data for a 'perfect reflector' (that is, a material that reflects 100 per cent of radiation across the range of wavelengths to be considered).

The spectrum of the sample is then measured and compared – wavelength by wavelength – with the data from the reference sample to give a spectrum that is independent of the light source. It is generally unnecessary to record a reference spectrum before every measurement, so the spectrum gathered at the beginning of the measurement session can be compared to all the subsequent sample spectra, providing the experimental conditions have not changed.

If quantitative data are required, a so-called **dark calibration** is also performed before each session. This accounts for the signal 'noise' generated by the photomultiplier tube or CCD in the absence of any light and is generally effected by blocking all light from reaching the detector.

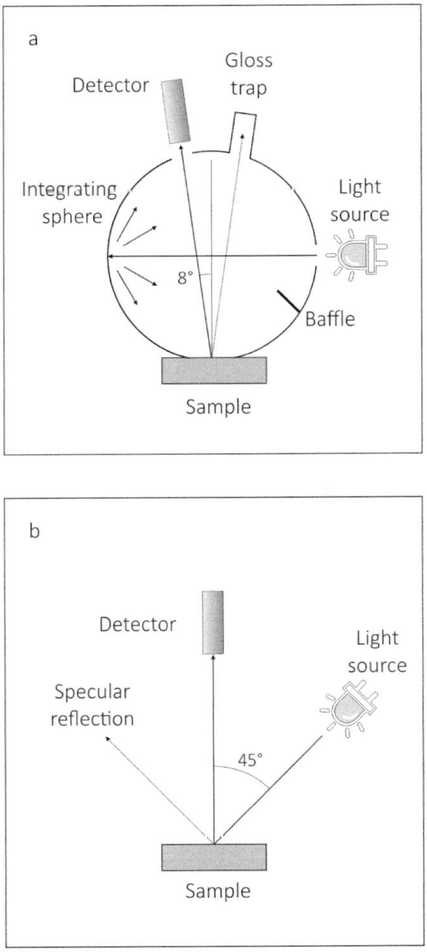

Figure 3.5 Two of the standard measurement geometries for reflectance recommended by the CIE: (a) d/8; and (b) 45/0

3.3.4 Geometry

If measurements are to be repeatable, then it is important to define the **geometry of measurement**. We saw in Chapter 1 that when light strikes smooth surfaces, specular reflection (gloss) is visible at certain angles. In measuring colour, we generally wish to discount the effect of gloss, although later in this chapter we will look at how the latter can itself be measured. This is generally accomplished by choosing a geometry for illumination and detection that either avoids causing specular reflection or ensures that if it is present, it does not reach the detector.

Helpfully, the CIE recommends standard illumination and detection geometries for colour measurement (CIE 2004a). Two of the most common geometries for gloss exclusion are termed d/8 and 45/0. In the first of these configurations, the sample is illuminated with diffuse light (the 'd' in the d/8 descriptor), produced by lighting the white inner surface of a so-called integrating sphere. The detector is oriented at 8° to the perpendicular of the sample (Figure 3.5a). This offset from the perpendicular permits the inclusion of a 'gloss trap' within the sphere, generally a tube whose interior is painted matt black to absorb all light. The sphere also includes a baffle to stop light from the source reaching the sample surface directly (Figure 3.5a).

In the second configuration (45/0), the sample is illuminated at an angle of 45° to its surface and the reflected radiation is detected perpendicular to this surface (i.e., at 0°) (Figure 3.5b). Although specular reflection – shown by the dashed line – may be generated at 45° to the surface if the object is glossy, it will not be detected. In some circumstances, the illumination and detection angles are reversed, termed 0/45 (Hunt and Pointer 2011: 104–105).

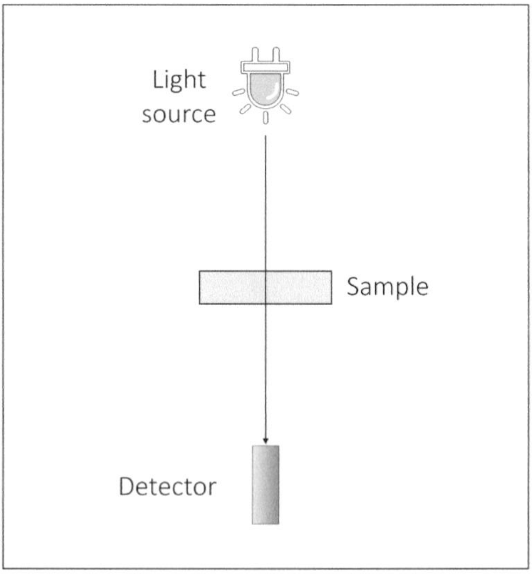

Figure 3.6 The standard geometry for transmittance measurements, with the source and detector aligned on opposite sides of the sample

For transmittance or absorbance measurements, it is usual for the light source and detector to be placed on the same axis, with the sample (or more rarely, object) placed between them (Figure 3.6).

3.3.5 Field of view

Another factor that needs to be considered is the field of view of the instrument, that is, the area on the surface of the sample over which the measurement is carried out. Unfortunately, there are no standards covering this and what constitutes an appropriate measuring area will depend on the texture of the object's surface, its colour homogeneity and the reasons for which the colour measurements are being made.

For a flat, smooth surface with a very homogeneous colour, the field of view chosen is less important. If the surface is smooth but the colour is inhomogeneous, then either a larger field of view may be selected to give a colour measurement over a representative area of the surface, or an attempt is made to focus on a smaller area of uniform colour. If the purpose of measurement is to look at subsequent changes in colour, the latter approach makes it more challenging to ensure that future measurements are made on exactly the same area (see Section 6.3).

For textured or curved surfaces, the central issue is that the controlled geometry we have defined for our instrument no longer applies. Sections of the surface will present at different angles to the light source or detector and the methods employed to eliminate gloss may not be effective. Under these circumstances it is generally better to employ a diffuse illuminant and to measure a larger area to average out the effect of the orientation of different portions of the surface. An instance where this approach may be applied is in the measurement of textile samples.

In practice, the field of view chosen, and indeed the measurement geometry, may be limited by the practicalities of instrument availability.

3.3.6 Types of spectrometer

We typically encounter two types of spectrometer: those that are termed 'benchtop', which, as the name suggests, are usually based on the laboratory bench and to which samples are brought, and portable or hand-held spectrometers that can be used to make measurements either in the laboratory or *in situ*.

Benchtop transmittance spectrometers are most useful for looking at detached samples, for example, glasses and transparent plastics, or the properties of liquids, which are held in the light path within containers with flat and parallel sides (cuvettes) that do not absorb in the visible region. For reflectance spectrometry, benchtop devices typically use an integrating sphere, and it is at the aperture of this sphere that the sample is placed. While it may be possible to bring certain collection objects to the port of a benchtop spectrometer, they are used mainly to assess detached samples.

Some portable and hand-held devices also include a small integrating sphere, particularly those that also serve as colorimeters, while others employ the 45/0 measuring geometry. The latter instruments commonly use fibre-optic guides to channel light from a stabilised light source to a small measuring head placed close to, or on, the surface of the object. Radiation reflected from the object is collected by the measuring head and passed through a second (often collinear) fibre-optic guide to the spectrometer for analysis. The measuring head often allows the positions of the fibre-optic guides to be adjusted to give different illumination and detection geometries, including the 45/0 geometry recommended by the CIE.

The technique described above, termed **fibre-optic reflectance spectrometry (FORS)**, is widely used in the cultural heritage field, not only to measure colour but – usually by extending the range of wavelengths into the ultraviolet and infrared regions – to undertake material identification by comparison with published libraries of spectra (Picollo et al. 2000; Cosentino 2014). The wavelength – or spectral – resolution of FORS devices is often quite high, 1 nm or less, in order to capture the fine details of the absorptions that can help to characterise particular molecules. Because the measuring head is often small, light and manoeuvrable, FORS instruments can be taken into museums to measure colour *in situ* and can make measurements at difficult-to-access points on the surface of an object.

3.3.7 Spectroradiometry

We have been focusing on the properties of objects and the way in which they reflect, transmit or absorb light, but in Chapter 1 we saw that light sources can be defined by their spectra, see Figures 1.4 and 1.12. In these spectra the power of radiation emitted is plotted against the wavelength; the resultant curve is termed the **spectral power distribution** of the light source. Such measurements are made with an instrument termed a **spectroradiometer**. Light emitted by a source is gathered by a lens and focused onto a prism or grating before being analysed by a detector (Figure 3.7); these elements are similar to those found in spectrometers used for reflectance or transmittance measurement. Spectroradiometric measurement is often employed within the cultural heritage sector to examine the properties of light sources used to illuminate collections, perhaps to assess their appearance or likely damaging effects on fugitive materials.

We can also record spectra for colours created by emission in light art, such as those seen in Figure 2.3. As installations containing neon, fluorescent or other sources become more prevalent, measurement of the light emitted can identify the type of lamp used, help to monitor any changes in colour as the sources age and inform a good colour match when obsolete lamps need to be replaced with a new lighting technology (Eng, Preusser and Schaeffer 2016; Healy n.d.).

In emission spectra we generally report relative power, that is, the power at each wavelength relative to the maximum value at any wavelength across the visible spectrum; for example, the data in Figure 1.12 are relative to the strongest emission

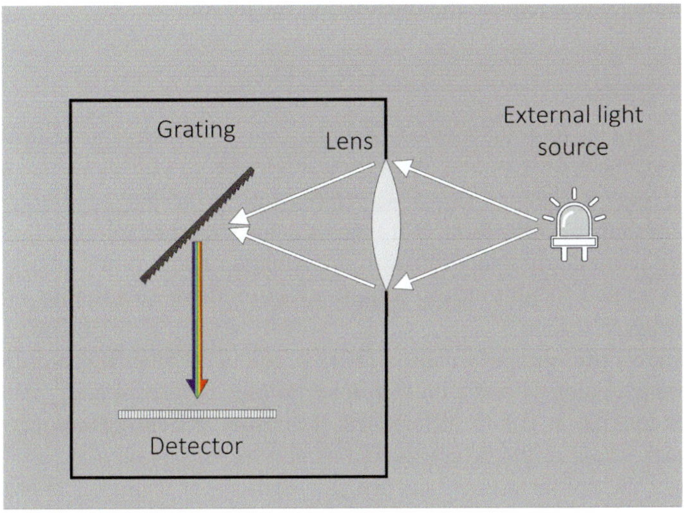

Figure 3.7 Schematic representation of the configuration of a spectroradiometer

at 545 nm, which is set to 100 per cent. If we wish to calculate and present the absolute power emitted by a source, the spectroradiometer needs to be calibrated by recording the spectrum of a reference source with a known spectral emission placed at a specific distance from the spectroradiometer.

3.3.8 Microspectrophotometry

A variation on the spectrometric methods we have encountered above is microspectrophotometry. It does not differ in principle from the methods applied to larger samples, but typically uses a bench or travelling microscope to deliver light to, and collect information from, the sample. In addition to studies of an object's surface or detached samples, it can also be applied to the examination of cross-sectional samples, where it may be possible to record colour information from individual layers or particles, depending on the resolution of the microscope. Transmittance microspectrophotometry on samples prepared as thin sections often produces clearer spectra than reflectance methods and has, for example, been used as a means to classify red lake pigment particles in paint sections (Kirby 1977; 1999).

Microfade testing (MFT) will be discussed in more detail in Section 6.6. Although its main function is to test the stability of materials to light exposure, it does so by recording reflectance spectra across a very small area of an object or sample surface (typically an area with a diameter of <0.5 mm). Some caution is needed if the equipment is used purely to measure colour, as the highly focused incident light beam is intended to cause colour change in less stable materials in a matter of minutes, and it may be necessary either to make very rapid measurements or – preferably – to attenuate the light output by the source.

3.4 Colorimetry

Spectra, as produced by the methods detailed above, contain a great deal of information about the reflection and absorption properties of objects, which are, as we have seen, related to the molecular structure of materials and can thus sometimes be useful in their identification. But, even for the experienced user, these spectra do not give particularly straightforward information about how colours appear to us and it is quite difficult to use spectra to give simple or quantitative information about changes in colour.

Colorimetry presents information about a colour as a set of co-ordinates that define it unambiguously and can, therefore, be used both to specify colour and as a basis for looking at colour differences. It is usual to define a colour with three co-ordinates. These might be considered analogous to the notations used to define a sample in the colour atlases that we saw earlier, for example, hue, value and chroma in the Munsell system.

Modern colorimetry relies on standardised colour co-ordinates defined by the CIE. These colour co-ordinates, or colour spaces, all have their origins in studies of the spectral sensitivity of the photopigments in the eye that led to the publication of a set of so-called colour matching functions by the CIE in 1931. These functions, shown in Figure 3.8, were derived experimentally by averaging the results across

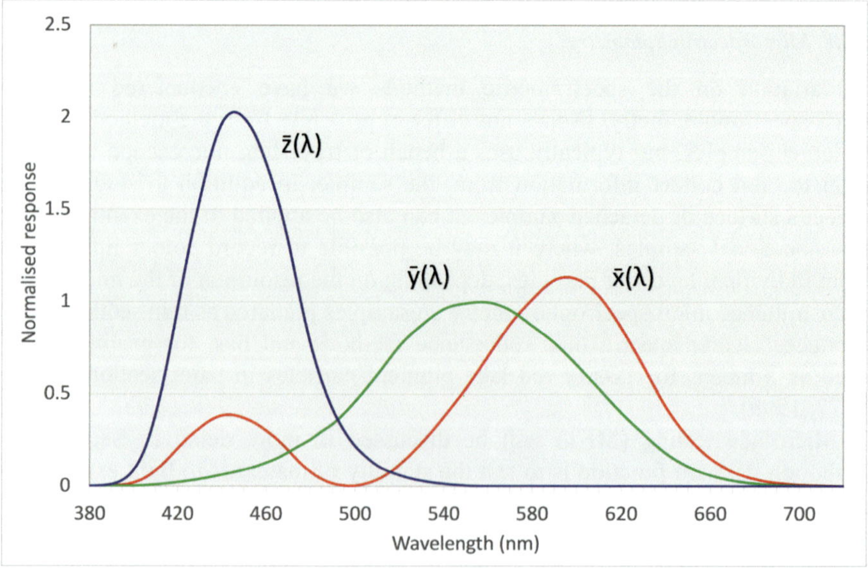

Figure 3.8 The colour-matching functions for the CIE standard colorimetric observer

a panel of observers as they matched colours that were displayed in one half of a 2° field of view by mixing monochromatic (single wavelength) red, green and blue lights in the other half of the field. The three functions shown in Figure 3.8 are represented by the symbols $\bar{x}$, $\bar{y}$ and $\bar{z}$ (read aloud as 'x bar', etc.), and are frequently referred to as the 2° observer functions (CIE 2004a).

These colour-matching functions allow us to convert spectral information into CIE colorimetric information. Although a considerable number of **CIE colour spaces** have been developed and used over nearly a century, we will only concern ourselves with two, which are the current CIE Lab system and the CIE xyz system, which was the first to be defined and is the basis for subsequent colour spaces and is still widely used as a method to present the chromaticity of colours. Details of other colour spaces can be found in standard colour texts mentioned in the Introduction.

3.4.1 The CIE xyz system

Starting with the colour-matching functions, we can define the way in which our visual system will respond at each wavelength across the visible region to a particular colour lit by a specific illuminant. Considering first the $\bar{x}$ matching function, if we represent the reflectance (or transmittance) of the colour at a wavelength λ as s(λ) and the power of the illuminant at that wavelength as E(λ), we can calculate an overall response across the visible range, X, using Equation 3.1. In the same way we can calculate Y and Z, which are the corresponding responses related to the $\bar{y}$ and $\bar{z}$ colour matching functions (Equations 3.2 and 3.3).

$$X = \sum_{380}^{700} s(\lambda)\ E(\lambda)\ \bar{x}(\lambda) \qquad \text{(Equation 3.1)}$$

$$Y = \sum_{380}^{700} s(\lambda)\ E(\lambda)\ \bar{y}(\lambda) \qquad \text{(Equation 3.2)}$$

$$Z = \sum_{380}^{700} s(\lambda)\ E(\lambda)\ \bar{z}(\lambda) \qquad \text{(Equation 3.3)}$$

It is important to note that the contribution of the illuminant – E(λ) – does not correspond to that of the illuminant used when the spectrum was measured. As we saw above, we generally report reflectance and transmittance data that are independent of the light source used in the spectrometer. The illuminant used in these equations is, rather, one of several **standard illuminants** defined by the CIE, of which those most frequently encountered in colorimetry are A, C, D50 and D65 (see *CIE standard illuminants* box).

CIE standard illuminants

The CIE defines a series of standard illuminants that are characterised by their emission spectra (spectral power distributions), but which do not necessarily correspond to realisable light sources (CIE 2009; 2010).

CIE illuminant A represents the standardised output of a tungsten incandescent lamp, the prevalent electric light source in 1931 when the standards were introduced; its spectrum is shown in Figure 1.4a.

CIE illuminants B and C were also introduced in 1931 and were designed to represent phases of daylight (B = noon sunlight and C = average daylight), in the visible region at least. Illuminant B is now rarely used, but CIE C was once a popular standard and still features in some colour calculations.

The members of the daylight (D) illuminant series, introduced in 1963, are defined in steps of 500 K from 5000 K upwards and termed D50, D55, D60, D65, etc. (CIE 2004a). The most common of these illuminants is CIE D65, which aims to present average daylight with a colour temperature of *c.*6500 K; its spectrum is shown in Figure 1.4b. CIE D50 is also used in some fields, particularly photography, and has recently been added to the schedule of CIE standard illuminants (CIE 2022a).

Tabulated data for these illuminants are available online from the CIE (CIE n.d.) or in standard colour texts (e.g., Hunt and Pointer 2011: 355–358).

The three responses – X, Y and Z – are termed the **tristimulus values** and it is from them that the xyz co-ordinates in the first of the colour spaces we will consider are derived, using Equations 3.4–3.6.[2]

$$x = \frac{X}{X + Y + Z} \quad \text{(Equation 3.4)}$$

$$y = \frac{Y}{X + Y + Z} \quad \text{(Equation 3.5)}$$

$$z = \frac{Z}{X + Y + Z} \quad \text{(Equation 3.6)}$$

Because the sum of x, y and z will always be one, it is usual only to quote the first two **chromaticity co-ordinates** (x and y), which are often presented on the **chromaticity diagram** illustrated in Figure 3.9. This diagram has been used

2 The use of $\bar{x}$, x, X – and their equivalents for y and z – can be confusing. $\bar{x}$, $\bar{y}$ and $\bar{z}$ are the colour matching functions, X, Y and Z are the tristimulus values and x, y and z the chromaticity co-ordinates.

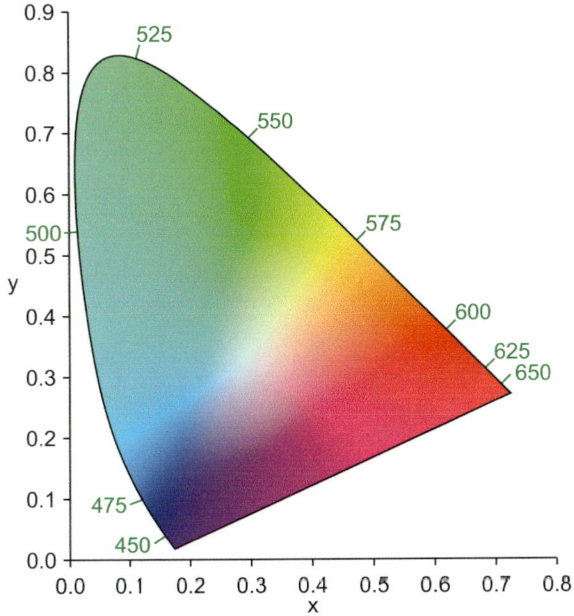

Figure 3.9 The CIE xy chromaticity diagram. The curved perimeter line represents the range of colours produced by single wavelengths of light. The straight line at the lower edge is the so-called 'purple line' joining the lowest and highest visible wavelengths

for a variety of purposes including showing the position of colours or groups of colours, plotting the locus of white light sources with different colour temperatures (see Section 3.5), and presenting the range of colours that can be displayed by computers, phones and TVs or be realised by a printer.

Although the third chromaticity co-ordinate (z) is generally omitted, x and y are frequently presented alongside Y, as Yxy data. As we saw in Chapter 1, the perceived lightness of a colour derives principally from the response of green cones, represented by the $\bar{y}$ matching function from which Y derives.

3.4.2 The CIE Lab system, also known as CIELAB 1976

One of the major drawbacks of the xy chromaticity system is that its colour space is highly non-uniform. That is, colours that appear to us to be just visibly different (we will look at how exactly this is defined later) are not always the same distance apart when plotted on the chromaticity diagram; pairs of very similar colours are widely separated in the green region but much closer to each other in blue or violet area (Figure 3.10). This deficiency becomes particularly important when it comes to looking at colour differences.

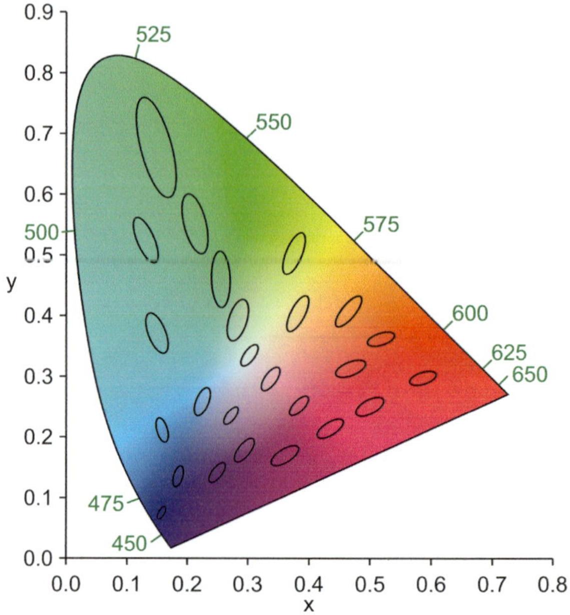

Figure 3.10 The CIE xy chromaticity diagram. The so-called MacAdam ellipses represent the extent of the colours that are indistinguishable from the central point of the ellipse, and they can be seen to vary in shape and size. Note that the ellipses are shown at 10 times actual size

In addition, although the terms x, y and z are very closely related to the stimulation of the three cones, we saw in Chapter 1 that it is Hering's opponent signal principle (rather than Young and Helmholtz's trichromatic theory) that is now generally accepted as a model for the human visual process.

Some of these shortcomings in the xy system are addressed in the CIE Lab colour space currently recommended by the CIE, first introduced in 1976 and with a history of use over 50 years (CIE 1978). The L*, a* and b* co-ordinates in this system correspond to the three opponents signals detailed in Chapter 1 and illustrated in Figure 1.10; the asterisks indicate that the values of L*, a* and b* have been calculated using the standard procedures recommended by the CIE.

The L* co-ordinate denotes lightness on a scale from 0 (pure black) to 100 (pure white). The a* co-ordinate indicates a position on a red–green opponent scale, with zero as a neutral point, positive values of a* denoting redness and negative values greenness. Finally, the b* co-ordinate presents yellowness (positive values) or blueness (negative values), with zero again indicating the neutral point on this yellow–blue opponent scale. We can present these three co-ordinates diagrammatically (Figure 3.11).

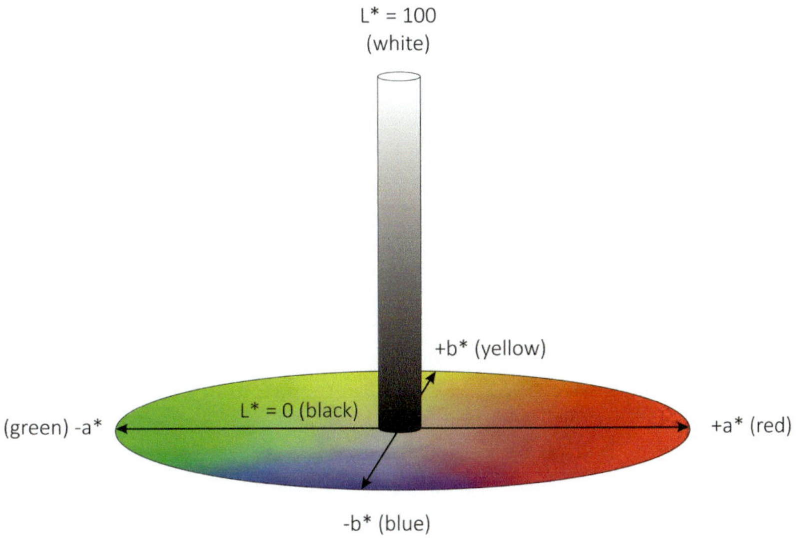

Figure 3.11 Schematic representation of the CIE Lab colour co-ordinate system

The L*, a* and b* values can be calculated from the tristimulus values X, Y and Z using Equations 3.7–3.9:

$$L^* = 116 f\left(\frac{Y}{Y_n}\right) - 16 \qquad \text{(Equation 3.7)}$$

$$a^* = 500\left[f\left(\frac{X}{X_n}\right) - f\left(\frac{Y}{Y_n}\right)\right] \qquad \text{(Equation 3.8)}$$

$$b^* = 200\left[f\left(\frac{Y}{Y_n}\right) - f\left(\frac{Z}{Z_n}\right)\right] \qquad \text{(Equation 3.9)}$$

These equations are also given in Appendix A, which includes details of the function *f*. The terms X_n, Y_n and Z_n used in these equations compensate to some extent for the different results that will be obtained if the illuminant used in the calculation is changed; they are the X, Y and Z values for a perfect white measured under that illuminant. The values of X_n, Y_n and Z_n given in Table 3.1 allow L*, a* and b* to be calculated from values of X, Y and Z for the four standard illuminants we encountered earlier – A, C, D50 and D65.

Table 3.1 Values of X_n, Y_n and Z_n used to calculate CIE Lab data for four CIE standard illuminants and two CIE observers

CIE standard illuminant	*CIE observer*	X_n	Y_n	Z_n
A	1931 2°	109.850	100	35.585
	1964 10°	111.114	100	35.200
C	1931 2°	98.074	100	118.232
	1964 10°	97.285	100	116.145
D50	1931 2°	96.422	100	82.521
	1964 10°	96.720	100	81.427
D65	1931 2°	95.047	100	108.883
	1964 10°	94.811	100	107.304

We can see that two different sets of figures are given in Table 3.1 for X_n, Y_n and Z_n for each illuminant, labelled '1931 2°' and '1964 10°'. The former refers to the narrow field of view used by observers in the experiments that led to the original 1931 **CIE colour matching functions**, while the latter derive from subsequent studies using a wider (10°) field of view. Most instruments and software programs allow the user to select the illuminant and observer conditions under which they wish to report CIE Lab data. Which is used is largely a matter of choice, although D65 data are more frequently cited. Consistency is probably more important and is vital when comparisons are to be made or colour differences calculated.

3.4.3 The CIE LCh system

The CIE data are sometimes expressed as lightness (L*), chroma (C*) and hue (h*), termed CIE LCh. The L* co-ordinate is identical to that in the CIE Lab space. As in the Munsell and other colour order systems, the hue is the 'nature' of the colour, here expressed as an angle with red at 0°, while the chroma is its saturation, indicating how far the colour lies from the neutral achromatic (white–grey–black) axis (Figure 3.12). The C* and h* co-ordinates can be derived from CIE a* and b* data as follows:

$$C^* = \sqrt{a^{*2} + b^{*2}} \qquad \text{(Equation 3.10)}$$

$$h^* = \arctan\left(\frac{b^*}{a^*}\right) \qquad \text{(Equation 3.11)}$$

We have seen that from its spectral information, any colour can be assigned an unambiguous set of CIE Lab co-ordinates and that we can even make a qualitative deduction about the nature of the colour from these co-ordinates. For example, CIE Lab data have been calculated for the two spectra shown in Figure 3.13.

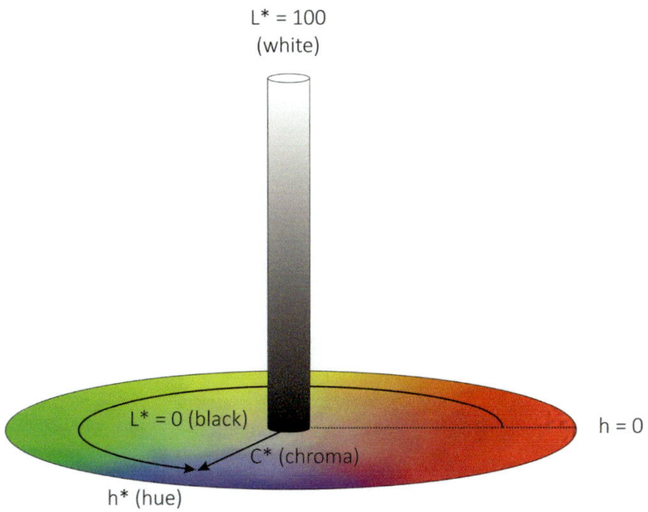

Figure 3.12 Schematic representation of the CIE LCh colour co-ordinate system

Colour 1 has an L* value close to the middle of the 0–100 range for L*, implying that it is moderately light, a large positive a* value, signifying redness and positive b* (yellowness), which suggests that this red is rather more orange than purple. Colour 2 has a somewhat lower L* value and is, therefore, a little darker. The small positive a* value indicates slight redness, but most significant is the large negative b* value, which identifies this colour as strongly blue. The spectra in Figure 3.13 are for the pigments chrome red (1) and azurite (2).

As the reduction of spectral information to a set of colour co-ordinates is modelled closely on our visual system, it might be no surprise that metamerism is a possibility. While, as we saw above, spectral information can be unambiguously assigned colour co-ordinates, the reverse transformation is not possible. A given set of CIE Lab colour co-ordinates cannot be used to generate an unambiguous spectrum, as more than one combination of surface colour and illuminant can produce the same colour co-ordinates. So, even though the CIE Lab data calculated for two materials with different spectra under a particular illuminant are identical, implying a visual match, when a second illuminant is used in the calculation, the CIE Lab data may differ, indicating a mismatch.

3.5 Colorimeters

Historically, one of the attractions of **colorimetry** was the possibility of obtaining colour co-ordinates for a sample directly, without the need for sometimes unwieldy spectrometers and the subsequent conversion of the spectral information to colorimetric data, which could be quite an undertaking in the pre-computer age.

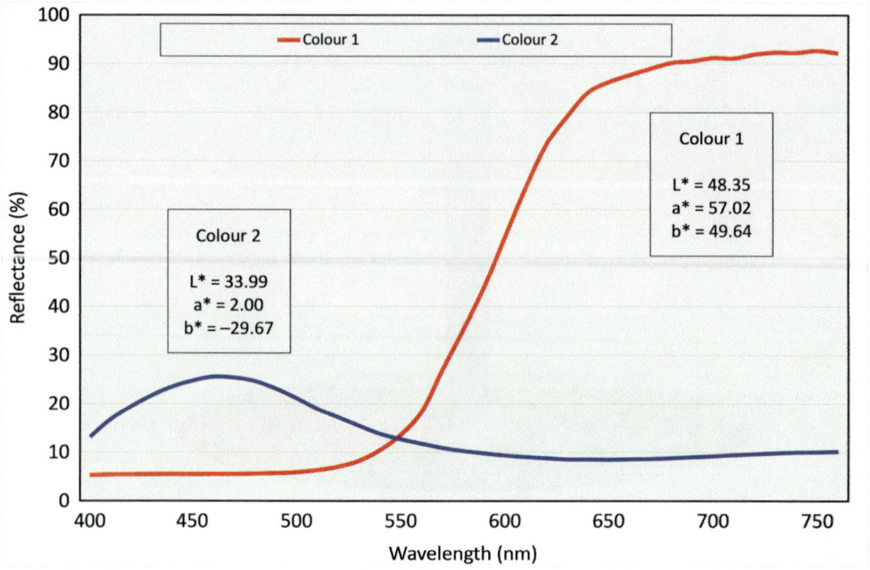

Figure 3.13 Reflectance spectra for two pigments and the CIE Lab data calculated from them; see text

The earliest **colorimeters** pre-date the definitions of standards for colorimetric measurement by the CIE. They were often developed for specific purposes, for example, the Lovibond colorimeter that was adapted by the National Gallery in London to make early colour measurements on paintings was originally intended to be used to assess the colour of beer (Rawlins 1936). As the data they produced were not expressed in terms of a standard colorimetric system, they instead relied on the same set of filters being present in every instrument in order to produce self-consistent results. These colorimeters also had the disadvantage that matches between the light passing through the sample and through the reference filters were made by the observer and were consequently subject to the quality and consistency of their judgement.

Once the colour spaces introduced by the CIE became established, colorimeters were developed in which the responses of sensors were matched as closely as possible to the CIE colour matching functions (see Figure 3.8), by applying customised filters to photocells that responded across the visible spectrum. Such colorimeters, which were widely used in the latter part of the twentieth century, removed the subjectivity of the human visual system from the measurements.

Advances in microchip technology mean that modern colorimeters tend to incorporate spectral sensors and microprocessors that convert the resulting spectra into colorimetric data directly. Although the wavelength resolution of these sensors is not always particularly high, typically 10 nm, this is quite adequate to allow the

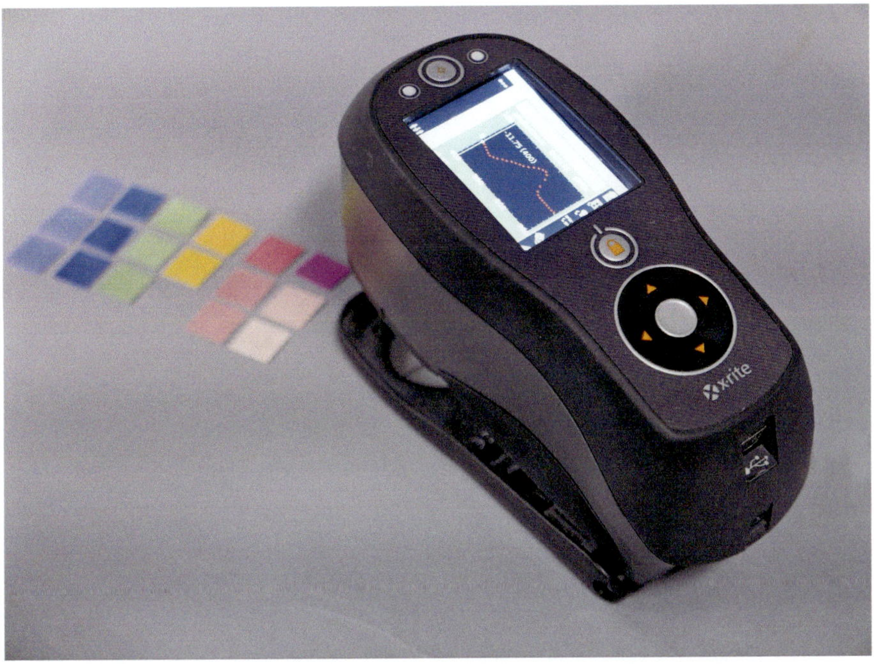

Figure 3.14 A typical modern colorimeter capable of displaying and storing CIE colorimetric data and the low-resolution spectral information from which these are calculated

information to be transformed into CIE Lab co-ordinates and an extensive range of other colour data. A modern colorimeter of this type is shown in Figure 3.14; note the large screen that can display colorimetric or – as here – spectral data.

It is worth noting that in the cultural heritage sector the term colorimetry is most frequently applied to the determination of colour co-ordinates by reflectance measurement. We can, however, generate colour co-ordinates from transmittance measurement (as was the case for the earliest colorimeters) or for light sources. In the latter, we tend to derive the colorimetric information from spectroradiometric data rather than measure it directly and it is common to present data for light sources on the CIE xy chromaticity diagram (Figure 3.15). Direct measurement, using a so-called **display colorimeter**, has long been applied to study the colour produced by TV, CRT, LCD and LED displays. The use of such displays in works of art and installations has increased from the late twentieth century and, when necessary, display colorimeters can be used to measure their emitted colour.

It is worth noting that the term colorimetry is sometimes used (by chemists in particular) to refer to a measurement – especially in liquids – at a single wavelength, generally in order to determine the concentration of a particular component in solution.

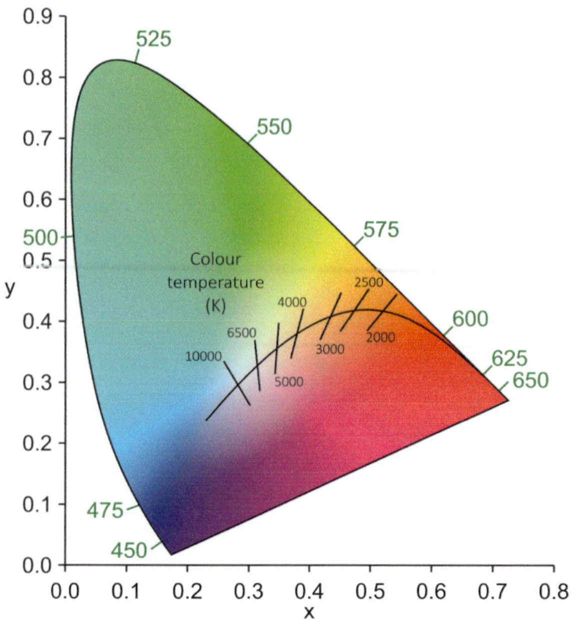

Figure 3.15 The CIE xy chromaticity diagram. The curved line represents the range of white light sources termed black body radiators, the approximate colour temperatures of which are indicated

Returning to the more common process of reflectance measurement, the extent to which colorimetry can produce quantitative, reproducible data will depend on several factors, many of which we have already encountered in our discussion of spectrometry. See the *Colorimetry checklist* box for a summary of this.

Colorimetry checklist

Parameters that should be included in an experimental record and reported alongside measurements (not all will be relevant to every instrument or software type):

- White calibration – and dark (black) calibration where possible
- Gloss condition (SCI or SCE)
- Aperture size (∅ mm)
- Standard illuminant used in calculation (D65, D50, C, A, fluorescent standards)
- Observer applied (CIE 1931 2° or CIE 1964 10°)
- Ultraviolet radiation (included or excluded)

3.5.1 Reflectance calibration

Before making a set of colorimetric measurements on objects or other samples, it is usual to calibrate the instrument by making a measurement of a white reference standard under the same conditions. As with spectrometry, this standard is typically a block of Spectralon with known reflectivity. With some instruments it is also possible to make a dark calibration by placing the measuring head of the colorimeter over the aperture of a dark trap (a cavity whose interior is painted matt black to absorb all light and comparable to the gloss trap encountered earlier). In some instruments the white standard – and perhaps the dark trap –will be incorporated into the baseplate on which the colorimeter is stored, facilitating alignment with the white and black standards.

The data collected during these white (and black) calibration steps will be stored automatically by the colorimeter and applied to correct readings made subsequently. It is good practice to recalibrate the instrument before each session and important to remember that recalibration is essential if any of the experimental parameters are changed, for example, the field of view of the instrument.

3.5.2 Geometry

Most portable reflectance colorimeters contain an integrating sphere to provide diffuse illumination and use the CIE d/8 geometry. If, as shown in Figure 3.5a, a gloss trap is present, the measurement of colour will be free from the influence of gloss; this measurement configuration is sometimes termed SCE (specular component excluded). On many instruments there is an option to measure the colour without eliminating the gloss component. This is achieved by closing the aperture to the gloss trap with a shutter whose inner surface is painted white so that light from the source may reflect from this region of the integrating sphere onto the object and thence to the detector. This measurement configuration is termed SCI (specular component included). It is more usual to find SCE measurements quoted and worth noting that the use of SCI colorimetry is not an efficient method of quantifying gloss, see Section 3.7 specifically on gloss measurement.

3.5.3 Field of view

As is the case for spectrometry, the area over which the measurement is made will depend on the nature of the surface under study (e.g., texture and colour homogeneity) and the reasons for measurement. Generally, colorimeters tend not to be used to measure very small areas or individual particles in the manner described above for spectrometers, particularly in cases where material identification is the goal.

The measurement area is usually circular, and many instruments offer a choice of interchangeable apertures to allow measurements to be made over different areas of a surface, for example, two widely available colorimeters offer a choice of 8 mm diameter and 3 or 4 mm diameter apertures. It is important to remember, however, that the colorimeter will need to be recalibrated each time the measurement area is changed, see the *Colorimetry checklist* box.

3.5.4 Colour co-ordinate calculation

The quality of the colour data reported by an instrument depends on the type of instrument. In older colorimeters the principal source of error lies with the quality of the match between the spectral response of the instrument's sensors and the CIE colour-matching functions used to specify standardised colour spaces. The same argument applies to smartphone-based apps that claim to produce colorimetric data; the match between the colour sensors in most cameras and the CIE colour-matching functions is poor, as discussed in more detail in Section 3.9. For colorimeters that capture spectral data, the wavelength resolution of the data and the accuracy with which this information is transformed into CIE co-ordinates are the determining factors.

As none of the above factors are under the control of the user once an instrument has been acquired, it is worth looking at how well the hardware and software elements of the colorimeter deliver colorimetric data before deciding on equipment.

3.5.5 Illuminants and observers

In Section 3.4 on colorimetry and in Table 3.1, we encountered different standards for measurement related to the illuminant and observer conditions used in the calculation of **CIE colour co-ordinates**. Modern colorimeters that convert spectral information to CIE Lab or other data can typically calculate these colour co-ordinates under one of several illuminants, using either the 2° or 10° CIE observer. As there are no absolute standards for these parameters (although some illuminants are more commonly applied than others), the most important consideration is to record which illuminant and observer have been applied in making a set of colour measurements and to be consistent in their use within a study or series of experiments.

3.5.6 Ultraviolet radiation

Depending on instrument type, the light sources used in measurement may emit **ultraviolet radiation** in addition to visible light. Some colorimeters incorporate both visible and ultraviolet sources, while in other cases, a single source (e.g., a xenon flash lamp) emits in both regions of the spectrum, but with an option to filter out the ultraviolet component if desired. It is usual to opt to exclude ultraviolet radiation, so that fluorescence generated in the visible region by excitation in the ultraviolet (see Section 2.1.10) does not interfere with colour measurement. There are, however, occasions when it may be advantageous not to exclude the ultraviolet component, for example, when studying fluorescent pigments and dyes, or the effect of optical brighteners in textiles, paper-based collections or photographs.

3.6 Colour difference

On many occasions we will measure colour in order to determine whether a change has occurred, or to set a benchmark against which future colour measurements can be compared.

3.6.1 Visual comparisons

When two colours are presented alongside one another, our visual system is capable of detecting small differences but, although there are those who claim to recollect colours accurately, we are generally not good at making colour comparisons based on memory. A visually-based process by which we can assess – and, to an extent, quantify – colour differences is with chips from colour atlases. For example, using the Munsell system we encountered earlier, we might match an area of an object to the chip bearing the notation 5R 5/8. If we suspect that after some time on display the colour has altered, we can perform the colour-matching process again, perhaps determining that the best match is now to the chip 5R 6/6. If we assume that the colour chips and our ability to make matches are unchanged, then we can say that the colour has changed – the hue is 5R (red) in both cases so has not changed but the value (lightness) has increased from 5 to 6 and the chroma (saturation) has decreased from 8 to 6.

While they are not generally as sensitive to small changes as instrumental methods, comparisons using colour atlases can be appropriate when equipment is not available, perhaps, as described in Section 3.1, for *in situ* or field studies, or where measurements are being made at a number of institutions that have colour atlases but lack comparable instrumentation.

3.6.2 Spectrometric comparisons

A comparison of the spectra for two colours, or for an area before and after a change is suspected to have occurred, can usefully reveal the nature of the difference in terms of the spectral regions affected and the likely alterations to the chromophore. If a peak in the spectrum that can be attributed to a particular component increases or diminishes during colour change, quantitative information about the rate of change can be accessed by measuring the absorbance, transmittance or reflectance at the peak wavelength.

It is less easy, however, to convey changes in the spectra as differences in the visual attributes of the colours. A general increase in reflectance or transmittance indicates that a colour is lighter, while a shift in an absorbance edge or peak implies a change of colour, but it is difficult to be more precise or to relate these changes to common expressions of colour, such as those articulated by the three opponent scales formalised in the CIE Lab system (light–dark, red–green and yellow–blue).

Accordingly, most colour difference measurements are made by first converting each spectrum into colorimetric data using the well-established methods detailed above and published by the CIE (CIE 1978; 2004a). Most spectrometers can now perform these calculations automatically, with many also able to determine the colour difference between the sample under investigation and the stored data for a reference colour.

3.6.3 Colorimetric comparisons

Once colorimetric data are available, either from direct colorimetric measurements or by converting spectral data, a colour difference can be calculated. Although there are various methods of expressing colour difference, by far the most common is that termed ΔE (in speech 'delta E', with the Greek capital letter delta indicating a change), which is based on the CIE Lab system and has been in use for nearly 50 years (CIE 1978).

There are several 'variants' of the **CIE colour difference (ΔE)**, the simplest of which (sometimes and usually in more recent publications expressed as ΔE_{76} to show it has been calculated using the 1976 CIE recommendations) is defined as the distance between the co-ordinates for the two colours when they are plotted in the CIE Lab colour space (Figure 3.16).

The colour difference ΔE_{76} can be represented mathematically as:

$$\Delta E_{76} = \sqrt{(L_1 - L_2)^2 + (a_1 - a_2)^2 + (b_1 - b_2)^2} \qquad \text{(Equation 3.12)}$$

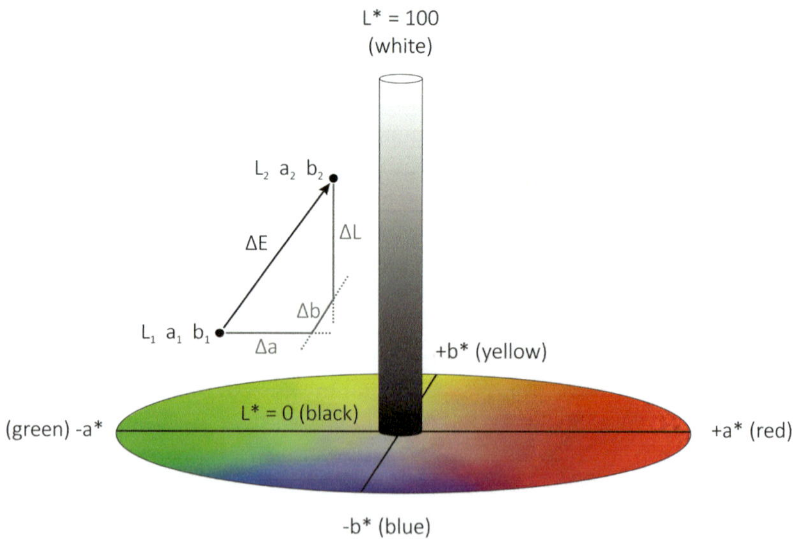

Figure 3.16 Schematic representation of the CIE Lab colour co-ordinate system showing the derivation of the ΔE_{76} colour difference measurement

where L_1, a_1 and b_1 are the CIE Lab values for the first colour and L_2, a_2 and b_2 the corresponding values for the colour with which we are making a comparison (Figure 3.16). Equation 3.12 is often expressed more simply, using ΔL, Δa and Δb to represent the differences in the three colour co-ordinates:

$$\Delta E_{76} = \sqrt{\Delta L^2 + \Delta a^2 + \Delta b^2} \qquad \text{(Equation 3.13)}$$

Changes in the three co-ordinates in the Lab system can themselves be helpful, allowing us to express a difference in terms of visual changes. For example, a negative ΔL indicates darkening and a positive Δb, yellowing.

The ΔE_{76} colour difference can also be expressed in terms of the CIE lightness, chroma and hue (LCh):

$$\Delta E_{76} = \sqrt{\Delta L^2 + \Delta C^2 + \Delta H^2} \qquad \text{(Equation 3.14)}$$

where

$$\Delta H = 2\sqrt{C_1 - C_2}\,\sin\left(\Delta h/2\right) \qquad \text{(Equation 3.15)}$$

In the same way that the CIE Lab system is intended to produce a more visually uniform colour space than some earlier methods, the definition of ΔE aimed to provide an indication of whether a colour change would be perceptible to an average viewer, with a colour difference (ΔE) of one intended to represent a 'just visible difference'. It was found, however, that while a colour difference of one was perfectly apparent when dealing with certain pairs of colours – particularly if they differed in hue – in other parts of the CIE Lab colour space pairs with much larger colour differences appeared identical.

To account for this issue, the simple colour difference formula used to calculate ΔE_{76} has been refined over the last 40 years so that a visually perceptible difference corresponds more closely to a particular value of ΔE, irrespective of the region of the CIE Lab colour space that is considered. The most significant of the various colour difference equations employed over this period are described briefly here, with more details available in standard colour texts, for example, Hunt and Pointer (2011: 61–64).

The first was developed by the colour measurement committee (CMC) of the Society of Dyers and Colourists and for this reason is known as the CMC equation (Clarke, McDonald and Rigg 1984). It was later modified to become the CIE 1994 colour difference equation, often abbreviated as ΔE_{94} (CIE 1995a). Both systems take as their starting point the CIE LCh space and apply a number of correction factors to account for its non-uniformity. Since, as mentioned below, ΔE_{CMC} and ΔE_{94} have now been largely replaced by ΔE_{00}, details of their derivation are not given, but can be found in, for example, Hunt and Pointer (2011: 61–65).

The measure of colour difference currently recommended by the CIE was developed from the CMC and CIE 1994 equations by refining the correction factors further. This metric is termed the CIE 2000 colour difference and is abbreviated as ΔE_{00} (CIE 2001; 2022b); see Appendix B for details of its calculation.

Although a measured difference of one ΔE_{00} unit will not always equate to a just visible difference, the CIE 2000 method provides a better, if not perfect, indication of how easy it will be to distinguish two colours and, as a standard that has been in use for over 20 years, is a well-established means of expressing colour difference compared to the shorter-lived methods it replaced.

The equations used to calculate ΔE_{CMC}, ΔE_{94} and ΔE_{00} are much more complex than the simple equation presented above for ΔE_{76} (which may be the reason that some researchers continue to use ΔE_{76}). However, modern spectrometers and colorimeters generally have software applications that calculate the colour difference between a measured and stored colour as ΔE_{00} units and, in many cases, as several additional colour difference metrics. In addition, there are online or freely downloadable converters or spreadsheets into which CIE Lab, CIE LCh, CIE XYZ or spectral data can be loaded in order to calculate ΔE_{00} or other colour difference metrics.

Despite the predominance of ΔE_{00}, it is good practice to indicate clearly which equation has been used to calculate reported or published colour difference data. Older published data for colour change may have been calculated using the ΔE_{76}, ΔE_{CMC} or ΔE_{94} methods, so it is important to take this into account when making comparisons between measurements made in different studies, and particularly frustrating when data are reported simply in ΔE units, with no indication of the equation used in their calculation.

3.6.4 Just visible differences

We have already encountered the concept of a just visible difference in colour – which is sometimes also termed a **just noticeable difference (JND)**[3] – and seen that methods of calculating colour difference have aimed to equate a numerical colour difference of one with a JND. While each successive iteration of the colour difference equation has advanced this aim, it is still not the case that a difference of less than one ΔE_{00} unit (in the current system) unequivocally indicates whether two colours appear identical.

In the cultural heritage sector, the concept of JNDs has been used to indicate that an object has visibly changed from its previous state, perhaps in response to an external environmental effect, such as exposure to light. It has, however, proved difficult to quantify a JND in terms of measured colour difference, with proposed values ranging from *c.*1 ΔE_{00} to 3–4 ΔE_{00}. Pretzel (2008) suggested that, under well-controlled experimental conditions, 1.5 ΔE_{00} may be a good compromise.

3 The JND is sometimes referred to as just noticeable fade (JNF), but as demonstrated in the text, not all colour change can be classed as fading, which correctly means loss of a colorant.

3.7 Gloss

We have, so far, regarded gloss as something of an annoyance – a phenomenon that potentially interferes with our ability to measure the intrinsic colour of an object. There will be occasions, however, on which we wish to measure the glossiness of an object and – in particular – to see if the degree of gloss changes, perhaps as a material ages or as a result of a conservation treatment. For example, when applying consolidants or fixatives to matt surfaces, such as pastel or charcoal drawings, we probably wish to ensure that greater physical stability does not come at the expense of an unintended more glossy appearance.

We saw earlier that the condition for gloss (specular reflection) is when the angle of incidence of the illumination and the angle of observation or measurement are equal. In spectrometers and colorimeters we carefully avoid this geometry, for example, by using a gloss trap. In designing a gloss meter, however, we need to set our illuminant and detector at the same angle with respect to the surface of the object (Figure 3.17).

Gloss meters incorporate a collimated light source that illuminates the sample at a particular angle. In instruments where the angle of illumination is fixed, measurements are usually made only at 60°, as shown in Figure 3.17, with the detector fixed at the same angle. This is the angle recommended for general gloss measurements (Nadal, Early and Thompson 2006). Other gloss meters allow the angle of the illuminant and detector to be changed, typically to allow measurements to be made at 20°, 60° and 85°.

Gloss is generally measured in 'gloss units' (abbreviated to GUs or sometimes Gs). The gloss of objects is referenced to that of a highly polished black glass

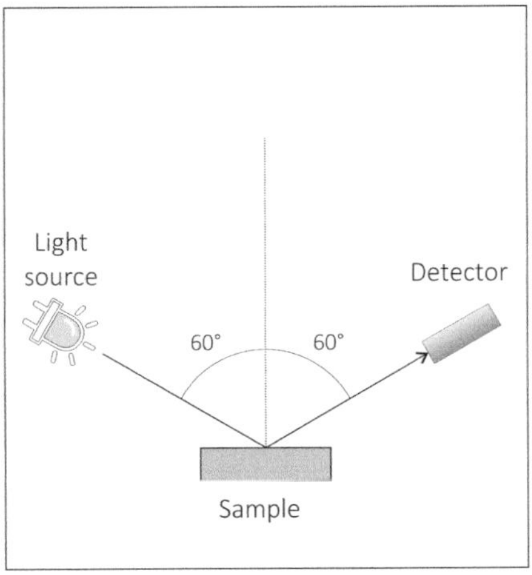

Figure 3.17 Recommended geometry for gloss measurements, with the source and detector positioned at 60° from the normal on opposite sides of the sample

standard, which is considered to have a value of 100 GU. Gloss meters are usually calibrated before a series of readings is made using such a standard.

When gloss is measured using a 60° angle of illumination, objects are sometimes classified as low gloss (<10 GU), medium gloss (10–70 GU) or high gloss (>70 GU). If an instrument with a variable illumination angle is available, it is recommended that high gloss is better measured using 20° illumination and low gloss using 85° illumination. It should be noted, however, that gloss measurements made with different angles of illumination will return different GU values. If comparisons (between samples or for the same material over time) are required, then the same angle should be applied.

Some newer models of colorimeter allow gloss (generally only at 60°) to be measured. While these colorimeters may be more expensive, they potentially replace two instruments.

3.8 Yellowness and whiteness indices

As we will see in Chapters 5 and 6, an ongoing concern in conservation is the potential degradation of plastics, organic coatings and conservation materials such as adhesives, a process that often manifests as yellowing. These changes can, of course, be measured using standard colour difference metrics such as ΔE_{00} or Δb^* (the yellow–blue opponent scale in the CIE Lab system), but it is also common practice to report such alterations using a Yellowness Index (YI).

The method standardised by the ASTM uses CIE XYZ data from colorimetry or spectrometry to derive the ASTM E-313 Yellowness Index (ASTM 2005):

$$YI = 100\frac{(C_X X - C_Z Z)}{Y} \quad \text{(Equation 3.16)}$$

The terms C_X and C_Z are shown in Table 3.2 for the two standard illuminants (C and D65) for which the index is defined.

In order to assess the potential discoloration of white materials, for example, paper and white plastics, the same ASTM standard (ASTM E-313) also defines a Whiteness Index (WI) that can again be calculated from CIE data (Yxy in this instance):

Table 3.2 Values of C_X and C_Z used to calculate the ASTM E-313 Yellowness Index for two CIE standard illuminants and two CIE observers

CIE standard illuminant	*CIE observer*	C_X	C_Z
C	1931 2°	1.2769	1.0592
	1964 10°	1.2871	1.0781
D65	1931 2°	1.2985	1.1335
	1964 10°	1.3013	1.1498

Table 3.3 Values of x_n and y_n used to calculate the ASTM E-313 Whiteness Index for three CIE standard illuminants and two CIE observers

CIE standard illuminant	*CIE observer*	x_n	y_n
C	1931 2°	0.3101	0.3161
	1964 10°	0.3104	0.3191
D50	1931 2°	0.3457	0.3585
	1964 10°	0.3477	0.3595
D65	1931 2°	0.3127	0.3290
	1964 10°	0.3138	0.3310

$$WI = Y + 800(x_n - x) + 1700(y_n - y) \quad \text{(Equation 3.17)}$$

The terms x_n and y_n are shown in Table 3.3, on this occasion for the three CIE standard illuminants recognised in the calculation of the WI.

3.9 Imaging colorimetry and spectrometry

The spectrometric and colorimetric methods discussed above provide single readings on a discrete area of a sample or object, but there are circumstances where it would be helpful to have information about colour across a wider area, for example, we might want to map the presence of a particular material across a painted surface.

A detailed description of imaging systems is beyond this short introduction to colour but there are essentially two ways of making such images. First, a point sensor can be scanned across the surface of the object, making sequential measurements. However, since digital cameras were first developed, there has also been interest in using them to make a grid of colour measurements, each corresponding to a pixel in the camera array.

Each technique has its advantages and disadvantages, which have changed somewhat as the associated technologies have developed. The greatest attraction of using a digital camera is speed, as the readings from each pixel are made concurrently. It may be necessary to make more than one image to cover the surface of an object at sufficiently high spatial resolution, but camera resolutions continue to improve (as we know from mobile phone technology). The disadvantages – if we are to make quantitative measurements – arise principally from two factors: the need to illuminate the object uniformly at a controlled angle and the aberration caused by the camera lens. In a scanning point system, the optics and illumination system tend to move with the sensor, removing the need to correct for uneven lighting with changing angle or lens effects. Slow scanning speeds and limited areas of coverage caused problems for early point sensor systems, but both are now much improved.

Whichever method is used, we need to consider how to extract colour information from camera or point sensor signals. Most cameras are designed for the photographic market and aim to produce pleasing rather than accurate or quantitative colour information. Such cameras rely on sensors coated so that they respond in the red (R), green (G) and blue (B) regions of the visible spectrum, the combined signals from which determine the colour at a given point in the image. The spectral responses of the RGB sensors vary between camera manufacturers and are not generally close matches for either the cone cells in the retina (see Figure 1.3) or the CIE colour matching functions (see Figure 3.8).

Mathematical transforms from camera RGB to either CIE XYZ or CIE Lab have been developed, but the colour accuracy across a range of colours remains low, with errors many times larger than a JND (Liang, Saunders and Cupitt 2005). Better colour accuracy requires a greater number of bands. One approach, developed by Berns (Imai, Berns and Tzeng 2000) is to use a standard colour (RGB) camera in combination with two interchangeable filters, which produces six-band images that have been used successfully in several applications, including the visualisation of changes to the appearance of objects, described in Section 6.2. A variation on this approach is to use a colour (RGB) camera with a set of narrow band LED lights. Compared to the RGB camera alone, the filter set roughly doubles the colour accuracy while the LED system nearly doubles the accuracy again (Kuzio and Farnand 2022: Table 2).

Alternatively, a camera that is sensitive across the whole visible spectrum can be combined with a set of bandpass filters to provide spectrometric information, albeit with rather low spectral resolution. Two such systems, developed within European projects, used seven or ten filters in the visible range. When the spectral information from these systems (VASARI and CRISATEL, respectively) was transformed into colorimetric data, the error in measurement was around the level of a JND (Liang, Saunders and Cupitt 2005).

While camera systems have largely been used to produce colorimetric data, the majority of scanning systems in current use produce spectral information. They employ sensors that respond across the visible range, with radiation reflected from the object surface divided into many discrete spectral bands using a grating in the light path, which replaces the bandpass filters used in earlier systems, and can offer a spectral resolution of 1–2 nm (Delaney et al. 2018). This is sufficient spectral resolution to allow material identification, particularly when the range of the instrument extends beyond the visible into the ultraviolet and infrared regions (Delaney and Dooley 2022). With appropriate calibration, accurate colour can also be derived from this spectral information.

The term 'multispectral' has often been applied to systems that used a series of bandpass filters, but this is now more frequently used to refer to the collection of several separate images in different spectral bands. Current techniques with very high spectral resolution are generally referred to as hyperspectral imaging or imaging spectroscopy. See Section 6.2 for some applications of these techniques.

3.10 Contact or non-contact measurement

While surface colour and gloss measurement are essentially non-destructive and non-invasive processes, they are not always non-contact. To maintain the correct geometry and to ensure that all ambient (surrounding) light is excluded, most measurements are conducted with the instrument in contact with the surface under investigation. This can present issues for less robust objects with surfaces that are friable or easily marked, and of course precludes measurements on drying paint or coatings.

In these cases, imaging methods provide a non-contact option, as do techniques that have been designed specifically to allow *in situ* non-contact measurement. These include FORS and spectrometers that were custom-designed to measure the surface of fragile objects (e.g., Bullock 1978). Another possibility for non-contact measurement is to use a spectroradiometer focused onto a very small area of an object, calibrated by making a reference measurement of a white standard under the same conditions.

4 Why colours change

We saw in Chapter 2 that colour arises either from the interaction of the physical structure of materials with light or by the selective absorption of light by atoms or molecules, i.e., as a result of their specific chemical composition. We might, for convenience and simplicity, refer to these as physical colours and chemical colours respectively.

Although changes or losses of colour are due to alterations in the physical or chemical properties of materials, it is not necessarily the case that chemical colours will be modified by chemical changes and physical colours by changes to their physical structure, as we will see in the sections that follow.

Some colour changes occur as the result of biological activity, but these are largely due to chemical reactions triggered or catalysed by compounds in the organisms responsible.

4.1 Chemical changes

The majority of the colour changes that we are likely to encounter – in the cultural heritage field at least – are due to chemical reactions within colorant or other molecules (intrinsic reactions) or, more frequently, of those materials with their environment (extrinsic reactions).

4.1.1 Activation energy

We think of certain colours as being prone to change, but the rates of many of the reactions described in the sections that follow are very low at ambient temperatures. These reactions will not proceed at an appreciable rate unless further energy is input into the system; the energy required to initiate reactions is termed the **activation energy** (sometimes denoted by E_a: see *Activation energy* box).

DOI: 10.4324/9781003397182-5

Activation energy

The activation energy is the minimum amount of energy necessary for a reaction to proceed. The concept is often presented diagrammatically, as below. Although the products of this reaction have a lower energy than the reactants, the end point can only be reached when sufficient energy is provided (E_a) to pass through the higher energy intermediate state.

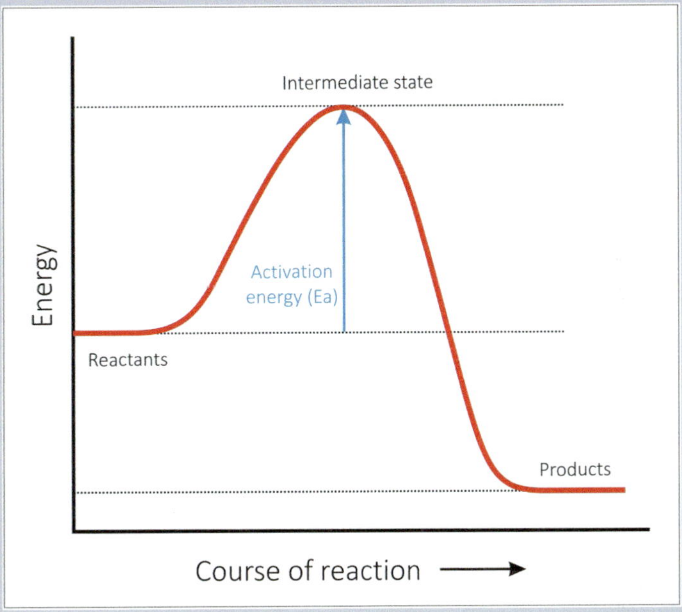

4.1.1.1 Light and ultraviolet radiation

Colour change is often associated with the action of light on materials and it is a paradox that the differential absorption of light that creates colour can also contribute to its change or loss. Furthermore, radiation beyond the visible region (not involved in producing colour) can also supply the energy needed to initiate or accelerate reactions that lead to changes in colour. Of particular concern will be the presence of ultraviolet radiation, which, as we saw in Chapter 1, has higher energy than visible light and is present to a greater or lesser extent in the radiation emitted by many white light sources (Figure 1.4), unless an ultraviolet filter is fitted to absorb it.

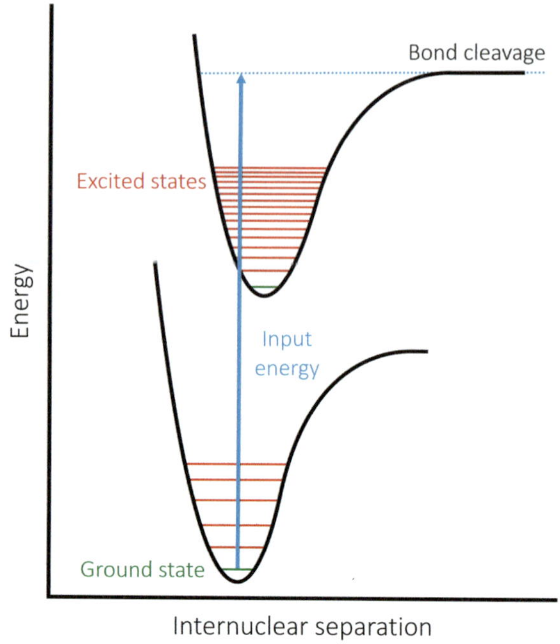

Figure 4.1 The excitation of a molecule from its ground state to an excited energy state at which bond cleavage (bond breaking) occurs

The energy associated with light and ultraviolet radiation can instigate both intrinsic and extrinsic chemical changes in coloured materials. Some of the intrinsic changes described below arise when sufficient energy is absorbed from light or ultraviolet radiation to trigger the breaking of bonds within the molecule (Figure 4.1). Other changes result from the absorption of energy by a molecule to create an excited state that is more reactive than the molecule in its ground state (Figure 4.2). In some cases, it is the colorant molecule that is excited, and which goes on to react with another molecule, while in other instances it is a second molecule that is excited, and which then reacts with the colorant. As we will see later, a common route for the deterioration of organic colorants in paints, inks and dyes is through the production of excited oxygen species.

Light may thus provide sufficient energy to allow a change to proceed much more rapidly than would be the case in the dark, a difference that can sometimes be seen where part of an object has been shielded from light (Figure 4.3). Many of the reactions in the sections that follow – such as oxidation, hydrolysis and reactions with pollutant gases – are accelerated in this manner.

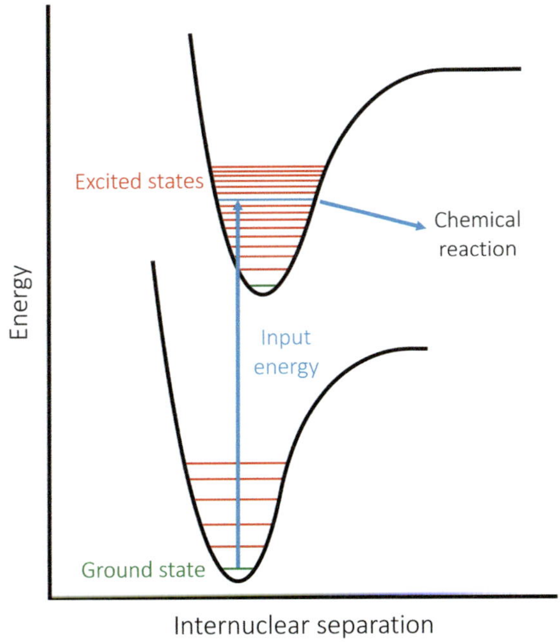

Figure 4.2 The excitation of a molecule from its ground state to an excited energy state that undergoes or instigates a chemical reaction

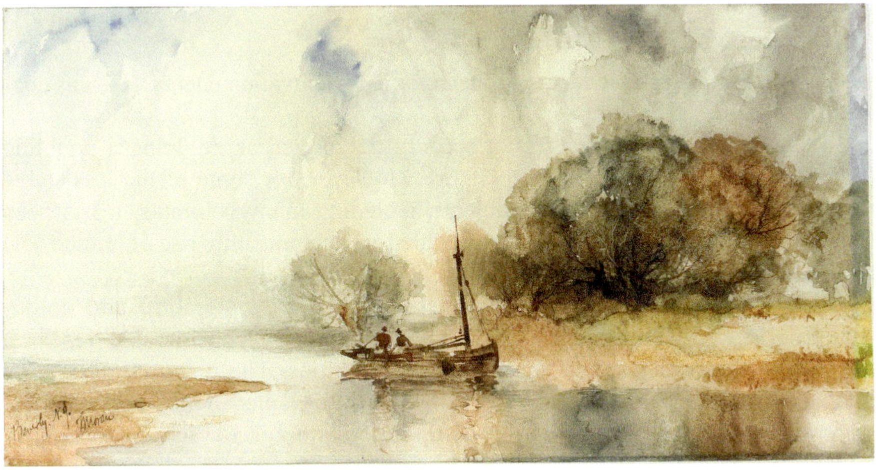

Figure 4.3 Watercolour painting of Beverly, New Jersey. Note the less-faded region along the right edge that has been protected from light by a frame or mount. [Source: Thomas Moran 1880–1889. 18.8 × 35.5 cm. Art Institute of Chicago, 1977.307. Olivia Shaler Swan Fund income.]

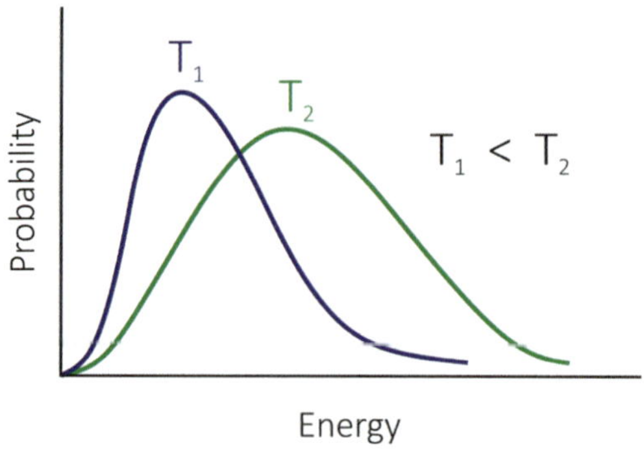

Figure 4.4 Maxwell-Boltzmann distributions at two temperatures; note that the distribution is shifted to higher energies and is broadened at the higher temperature

4.1.1.2 Heat

Another source of the activation energy needed for a reaction to proceed is an increase in temperature. As temperature increases, the average energy of molecules also increases. The distribution of energy at different temperatures is termed the Maxwell-Boltzmann distribution and is shown schematically for two different temperatures (T_1 and T_2) in Figure 4.4.

The higher energy can be linked to greater reactivity through the so-called Arrhenius equation (Equation 4.1), which relates the rate of a reaction to temperature:

$$k = Ae^{-Ea/RT} \qquad \text{(Equation 4.1)}$$

where k = reaction rate, A = Arrhenius factor, E_a = activation energy, R = gas constant, T = temperature.

In chemistry, this relationship is broadly interpreted as implying a doubling of rate for each 10°C rise in temperature. However, for some of the less stable materials we might encounter in objects, including many colorants, it has been found that the rate of reaction may be doubled by a temperature rise of around 5°C (Michalski 2002).

Most studies of the direct relationship between temperature and colour change relate to synthetic or natural polymers, with considerable research devoted to the behaviour of paper-based collections. The potential yellowing (and loss of strength) of drawings, prints, photographs, books and archival material have led to the use of low-temperature storage for some particularly vulnerable material.

For colorants, it has been more common to study the effect of temperature on light-induced degradation (e.g., Saunders and Kirby 2001). As expected, higher temperatures increase the rate of colour change, presumably either because the excited species generated by initial absorption of light react more quickly or because thermally excited species are produced that act as precursors for light-induced reactions.

4.1.2 Intrinsic reactions

Intrinsic reactions are less common than those involving the colorant molecule and a second compound. Intrinsic processes leading to colour change might be subdivided into three common categories.

- *Fragmentation of the colorant molecule*, which particularly affects those organic materials whose colour arises from delocalised bonding. When energy is absorbed, perhaps in the form of heat or light, then bonds may break within the colorant molecule, producing smaller fragments whose energy levels will differ from those in the original molecule, as a consequence of which, colour may be changed or lost.
- *Combination of the colorant molecule with itself*, which again principally affects organic materials. The linking of molecules, particularly long chain polymers, in which this phenomenon is usually termed cross-linking, can change the properties of the material., including its solubility, elasticity, fragility and colour (although changes to colour are less frequent). Such cross-linking can lead either to the loss of colour if the degree of delocalisation is reduced but, when conjugated systems are created, is one of the causes of yellowing.
- *Rearrangement of the atoms or bonds within molecules*, changing the energy levels and thus the colour. Most examples are in small inorganic molecules that can exist in more than one form, known as isomers, that have the same chemical composition, but different arrangements of the atoms within the molecule. Sometimes, one isomer converts to another slowly, but the process is often accelerated if energy is provided as heat or radiation. An example of a light-induced conversion is offered by the alteration of realgar, the orange form of arsenic(II) sulfide, to yellow pararealgar; both have the same chemical composition, but different molecular structures. Note, however, that pararealgar occurs naturally and has also been used as a pigment, so its presence does not always indicate that the colour of an object has changed.

For many years it was assumed that the darkening of the pigment vermilion (or the analogous natural mineral cinnabar) was due to the conversion of the red isomer of mercury sulfide (termed α-HgS) to a black form of the compound, metacinnabar. We saw in Chapter 2 that the band gap of 2.0 eV in vermilion lies behind its red colour, while metacinnabar has a band gap of 1.7 eV, which is outside the visible region. Recent research (discussed in later sections) has, however, shown that the process is much more complex, with metacinnabar only one of several possible products from vermilion degradation (Figure 4.5).

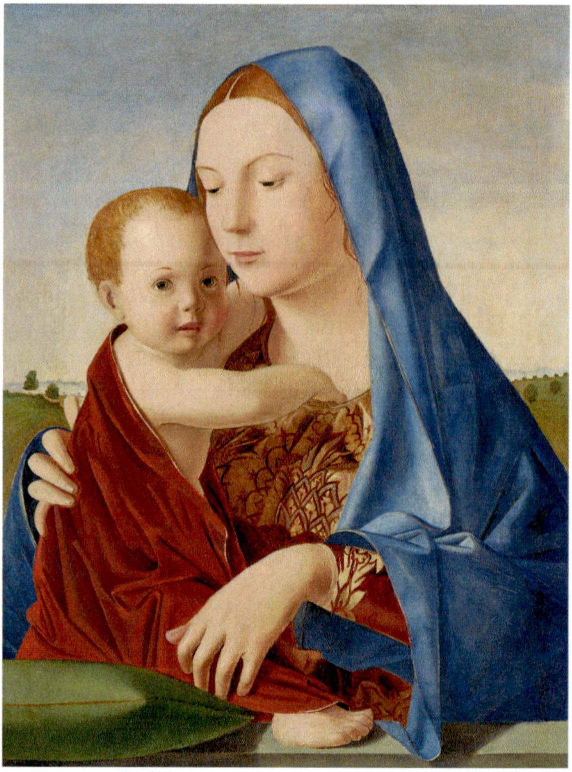

Figure 4.5 *Madonna and Child* by Antonello da Messina, showing discoloured vermilion in the red drapery wound around the child's body [Source: Oil and tempera on panel transferred from panel, *c.*1475. 58.1 × 43.2 cm. National Gallery of Art, Washington, DC,1937.1.30. Andrew W. Mellon Collection.]

4.1.3 Extrinsic reactions

Most colour changes arise from the chemical interaction of colorant molecules with other compounds. There are several ways to approach the different reactions associated with colour change. Here, we will look at the classes of material affected and introduce those reactions that alter their colour. For the purposes of this introduction to colour changes, three such categories are considered, metals, inorganic compounds and organic compounds.

4.1.3.1 Metals

Some metals, including gold, platinum and other so-called noble metals, are relatively inert so their colour and gloss are generally well preserved, even in extremely old objects or where they have been used as very thin foils to decorate manuscripts, paintings and other objects.

Many of the reactions that affect the colour of less inert metals arise from their oxidation. Oxidation can be defined either as the addition of oxygen to an element or compound, the loss of hydrogen or – as we saw in Chapter 2 – an increase in the oxidation state of an element, in this case, the metal. Conversely, reduction can imply a loss of oxygen, addition of hydrogen, or a decrease in oxidation state.

Direct reaction between materials and atmospheric oxygen is the simplest form of oxidation, but not necessarily the most common. For example, aluminium reacts readily with oxygen to form a patina of white aluminium oxide that, although altering the colour, also has the beneficial effect of protecting the aluminium metal beneath from further reaction – a process known as passivation. Here, aluminium has been oxidised in both senses of the term; it has combined with oxygen and its oxidation state has increased from Al^0 to Al^{3+}.

While corrosion processes for other metals – such as iron, zinc and copper – involve oxidation, they are also strongly influenced by the availability of water, either as liquid water or atmospheric moisture. For example, oxidation of iron or iron alloys in the presence of water leads to the formation of a familiar yellow–red–brown coating of rust, a mixture of hydrated iron(III) oxides and oxyhydroxides, the ratio of which will depend on the conditions of formation. Again, iron has been oxidised both by combining with oxygen and through an increase in oxidation state from Fe^0 to Fe^{3+}.

In addition to oxygen and moisture, a third component in the atmosphere can also contribute to the deterioration of metals. Among the final products of both zinc and copper corrosion are metal carbonates, created by reaction of the metal oxides or hydroxides produced in the initial stages of corrosion with atmospheric carbon dioxide. Zinc is corroded first to the hydroxide, which reacts with carbon dioxide to form white zinc carbonate, which forms a stable patina on the metal surface. Initially, copper metal is oxidised to reddish pink copper(I) oxide, which continues to react with oxygen to form black copper(II) oxide. Further reactions with water

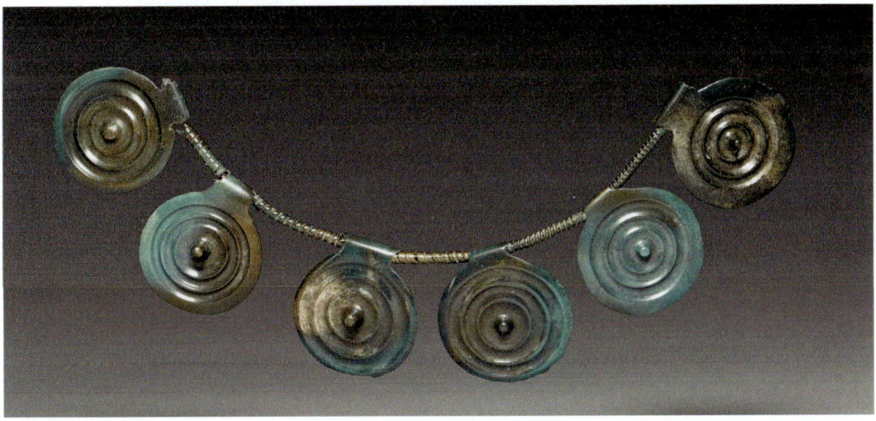

Figure 4.6 Green patination on the copper alloy elements of a necklace [Source: Hungary, Bronze Age, *c.*1400–1300 BCE. 4.2 x 4.8 x 0.8 cm. The Cleveland Museum of Art, 1992.68.a. Purchase from the J. H. Wade Fund.]

and atmospheric carbon dioxide yield green basic copper(II) carbonates, which are common components in the distinctive, stable, bluish–green patina on copper or copper alloy (e.g., bronze) surfaces (Figure 4.6).

As the reactions detailed above arise from interactions with components in the air – moisture, oxygen and carbon dioxide – it is difficult to prevent such corrosion or patination without the use of a very low relative humidity and/or oxygen-free environment.

Other reactions of metals that lead to colour change arise from interactions with gaseous pollutants. In the presence of moisture, sulfur oxides (SO_2 or SO_3, often collectively called SO_x) – gases that have a long association with the burning of coal and other fossil fuels – yield acidic solutions that may attack metals. In a well-documented

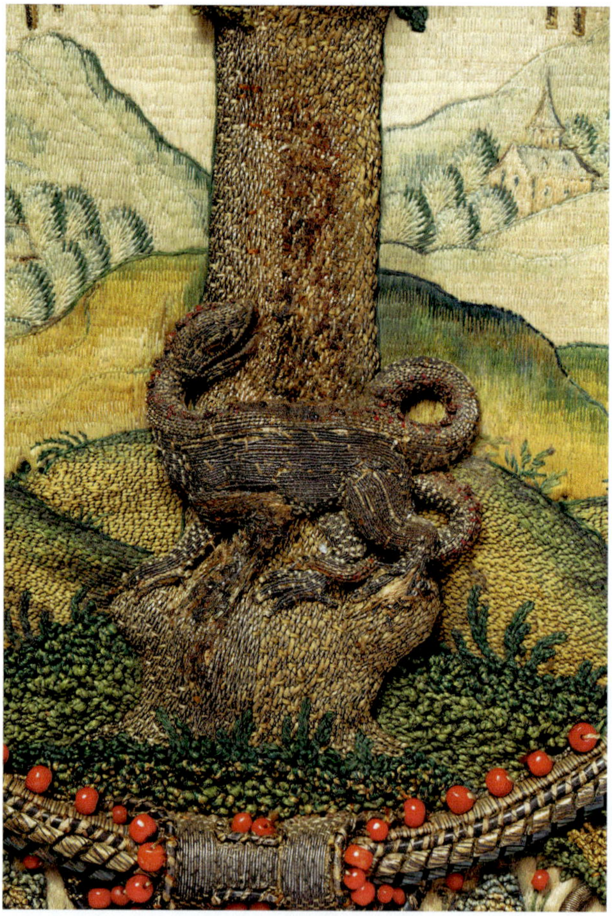

Figure 4.7 Detail of tarnished silver thread on an embroidered panel for a cabinet door [Source: France or Flanders, late sixteenth or early seventeenth century. 33 x 28 cm. Cooper Hewitt, Smithsonian Design Museum, 1959-144-1. Gift of Marian Hague.]

nineteenth-century example, the scientist Michael Faraday described a "copper apparatus" that was patinated with a "large body of sulfate of copper", i.e., blue–green copper(II) sulfate (Select Committee on the National Gallery 1850).

Reactions with reduced sulfur gases such as hydrogen sulfide (H_2S) or carbonyl sulfide (O=C=S) – in which sulfur is present as S^{2-} – are responsible for colour change in various metal surfaces, a phenomenon often termed 'tarnishing'. While black copper and nickel sulfides may be formed, tarnishing most commonly affects silver objects (Figure 4.7). Confusingly, such tarnishing of silver is sometimes termed oxidation: no oxygen is involved, the product of the reaction being black silver(I) sulfide (Ag_2S), although the oxidation state of silver increases during the reaction from Ag^0 to Ag^+.

Drawings executed in silverpoint frequently show a dulling of the metallic lines due to the formation of silver degradation products, the nature of which depends on the environment in which the drawing has been stored and exhibited. Oxidation to silver(I) oxide (Ag_2O) may account for some change, but where reduced sulfur gases are present, the most likely product is again black Ag_2S (Tanimoto and Verri 2009). In contrast, colour changes in leadpoint drawings begin with the oxidation of lead to lead(II) oxide, which can further react with carbon dioxide or sulfur dioxide (where present) to produce lead carbonate or sulfate (Bescoby, Rayner and Tanimoto 2010: 50–54).

Another common cause of colour change to metals is through their reaction with acidic gases generated by off-gassing from, or deterioration of, materials in their display or storage environment. For example, ethanoic (acetic) acid emitted by many types of wood reacts very rapidly with lead objects so that under certain conditions corrosion will be observed in just a few days. The lead ethanoate formed by this reaction reacts with atmospheric carbon dioxide to yield more stable lead carbonates, which are responsible for the white deposits on deteriorated lead objects.

4.1.3.2 Inorganic compounds

In common with metals, a number of coloured inorganic substances, many of them employed as pigments, are also affected by reactions with oxygen, water and atmospheric pollutants that alter their light absorption properties, changing or destroying their colour. The examples given below simplify somewhat the reactions involved, which are often part of more complex processes that are described in the articles cited.

4.1.3.3 Oxidation and reduction

Arsenic sulfides have been used as yellow or orange pigments since antiquity by European, African and Asian cultures. However, pararealgar, realgar and orpiment are all known to lose their yellow or orange colours on reaction with oxygen, fading as white arsenic(III) oxide (arsenolite: As_2O_3) is formed. As is often the case, this reaction proceeds more quickly when light or ultraviolet radiation is present (Keune et al. 2015; Broers et al. 2023). Many Egyptian Books of the Dead are

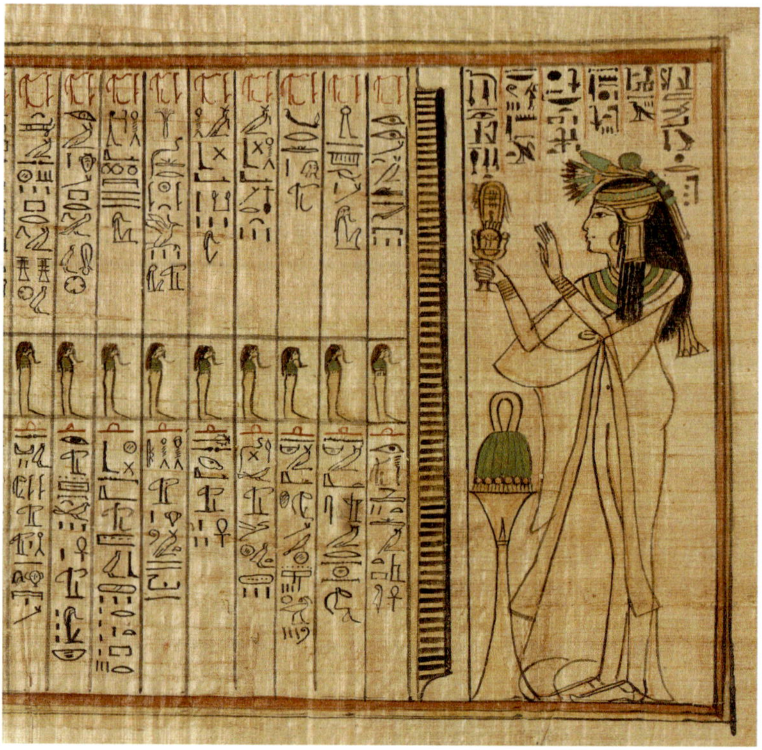

Figure 4.8 Detail from the papyrus scroll *The Book of the Dead for the Chantress of Amun Nauny*. The orange and yellow borders of such Books of the Dead are frequently painted using light-sensitive arsenic sulfide pigments, for example, realgar, orpiment or pararealgar. [Source: Deir el-Bahri, Egypt, *c.*1050 BCE. 521.5 x 35 cm. Metropolitan Museum of Art, 30.3.31. Rogers Fund, 1930.]

painted with borders comprising bands of orange realgar and yellow orpiment or pararealgar (Figure 4.8), the colour of which has been lost in those papyri that have been exposed to light more frequently (Daniels and Leach 2004).

Oxidation – probably exacerbated by light – is also implicated in the deterioration of cadmium-containing pigments. White cadmium sulfate ($CdSO_4$) has been found to be a degradation product of cadmium yellow (CdS) and cadmium orange or red ($CdS_{(1-x)}Se_x$), although further reactions yield both cadmium carbonate and oxalate, which are also white (Monico et al. 2022).

Lead white, a basic lead carbonate that has been found as a white pigment in both Classical and Chinese contexts since the fourth to third centuries BCE, is known to darken under certain circumstances. While the best-known route to such darkening is through reaction with reduced sulfide compounds including hydrogen sulfide, see below, lead white can also undergo oxidation to form brown lead(IV) oxide (plattnerite: PbO_2). The latter reaction has most often been reported in wall

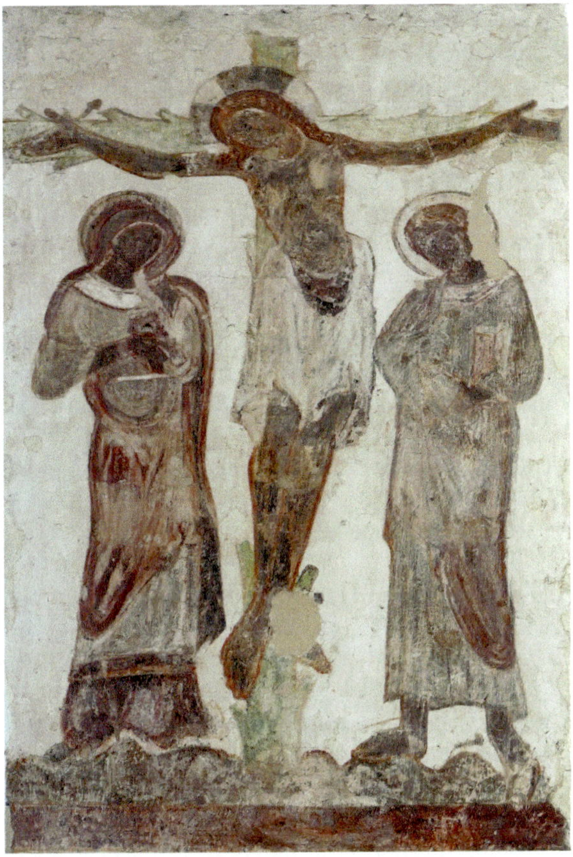

Figure 4.9 Late twelfth- or early thirteenth-century *Crucifixion* from the nave of St Albans Abbey. The flesh tones comprised red ochre and vermilion mixed with lead white that has darkened by oxidation to brown lead(IV) oxide

paintings (Figure 4.9), where it is thought that combinations of moisture, alkalinity or biological activity may facilitate the oxidative process (Giovannoni, Matteini and Moles 1990).

The pigment red lead can also be darkened by oxidation. As seen in Section 2.1.4, red lead contains both Pb^{2+} and Pb^{4+} ions. Oxidation of Pb^{2+} to Pb^{4+} removes the possibility for charge transfer between lead(IV) and lead(II) centres, which we saw in Chapter 2 is responsible for the red colour, and yields dark brown lead(IV) oxide (Aze and Vallet 2002).

Examples of colour changes due to reduction are a little less common. They arise mostly from changes in oxidation state, particularly those of metals, which – as we saw in Chapter 2 – can affect the energy levels and hence the absorption of light. We have already mentioned that the darkening of vermilion is not due to a single reaction and its changed colour may result, in part at least, from the reduction of vermilion, mercury(II) sulfide, to metallic mercury – a reduction in the oxidation state from Hg^{2+} to Hg^0.

The darkening of some copper-containing species is thought to result from the reduction of green copper(II) compounds to copper(I) species, which typically show a brown colour. Historically, this darkening was sometimes attributed to the deliberate use of a pre-prepared 'copper resinate' as a translucent green glaze, but the current view is that it occurs where certain copper(II)-containing compounds, particularly verdigris – copper acetate: $Cu(O_2CCH_3)_2$ – have already reacted with fatty or resin acids in the binder or an overlying varnish to form copper soaps, see Section 4.1.4.1 (Spring 2016).

There has also been considerable interest in the darkening of chrome yellow pigments, not least because they were used liberally by Vincent van Gogh. Although chrome yellow may contain one or more of a number of different compounds, their yellow colour derives from the presence of chromium(VI) ions. Lead chromate ($PbCrO_4$) itself seems to be relatively stable, but sulfate-containing versions of the pigment ($PbCrO_4 \cdot xPbSO_4$) are susceptible to reduction – exacerbated by light – to yield chromium(III) species, which are typically dark green (Monico et al. 2015).

Finally, the fading of Prussian blue, at least in the short term, is thought to be due to the reduction of iron(III) to iron(II), possibly by reaction with components in the associated binder or substrate. As we saw in Chapter 2, Prussian blue owes its colour to IVCT from Fe^{2+} to Fe^{3+}, so reduction of the latter will result in a loss of colour. The process is known to be promoted by light, but there is evidence that if partly faded Prussian blue is placed in the dark, reoxidation of Fe^{2+} to Fe^{3+} by atmospheric oxygen restores some of the colour, making it one of the few colour changes that is – at least partly – reversible. Another unusual feature is that because of this reoxidation step, the short-term fading of Prussian blue is one of the few reactions that is accelerated by the near-complete absence of oxygen (Lerwill et al. 2015); most reactions are much reduced in anoxic conditions. Over a longer period, a slower and irreversible oxidation of Fe^{2+} to Fe^{3+} also disrupts IVCT and leads to a permanent loss of colour (Kirby and Saunders 2004).

4.1.3.4 Pollutants

The reduced sulfur gases hydrogen sulfide H_2S and carbonyl sulfide O=C=S, which cause tarnishing in metals, can also affect some pigments. The best-known such reaction leads to the darkening of lead white in works of art on paper, through the formation of black lead(II) sulfide, PbS (Figure 4.10), a phenomenon that has also been noted in lead white containing ground and paint layers (Carlyle and Townsend 1990). Historically, poor standards of public hygiene and the prevalence of horse-drawn transport in cities resulted in significant levels of hydrogen sulfide that affected those works not protected within bound volumes. Other lead-containing pigments can also demonstrate blackening on exposure to reduced sulfur gases, including yellow massicot – lead(II) oxide – and red lead, although as

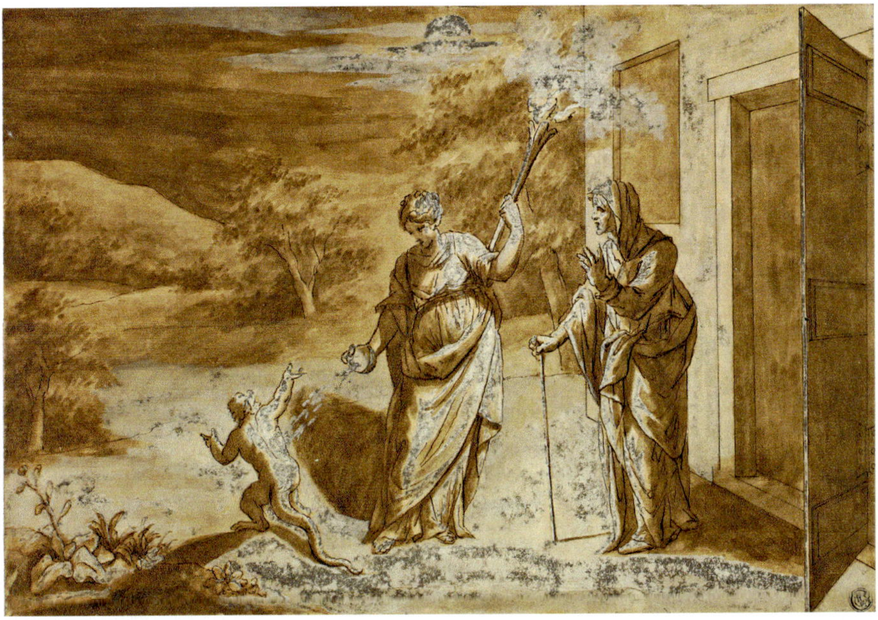

Figure 4.10 Pen and brown ink drawing of *Ceres Changing the Mocking Boy into a Lizard* with brown washes and lead white highlights, some of which have discoloured to dark grey [Source: Fontainebleau School, early seventeenth century. 26.3 × 37.7 cm. Art Institute of Chicago, 1922.302. The Leonora Hall Gurley Memorial Collection.]

mentioned earlier, darkening of the latter may also result from oxidation to brown lead(IV) oxide.

Other gaseous pollutants are implicated in oxidation reactions, including ozone and nitrogen dioxide, which occur naturally in the atmosphere and at greater concentrations in areas affected by photochemical smog. Although organic colorants are more susceptible to oxidation by these gases (see Section 4.1.3.10) those inorganic colorants that are susceptible to atmospheric oxidation – such as orpiment and realgar – change colour more quickly in their presence.

Earlier, we saw that in the presence of moisture, sulfur dioxide forms an acidic solution. While other oxides of nitrogen and sulfur (e.g., NO_2 and SO_3) can also yield acidic solutions, there are few reports of their reaction with inorganic colorants, although recent improvements in analytical methods have revealed the presence of sulfate-containing compounds. The reaction between dissolved sulfur oxides and calcium carbonate to yield calcium sulfate has, however, long been a concern in the conservation of wall paintings. Although both materials are white, their interconversion can lead to changes in the appearance of wall paintings, ranging from a white layer of calcium sulfate masking the surface colours to the loss of areas of paint.

A more serious concern can arise from acidic pollutants generated within buildings, display cases or storage enclosures; these reactions again proceed more quickly when moisture is present. We saw earlier that acetic acid reacts with certain metals to produce a surface coating of metal acetate. Acetic acid – and, to a lesser extent, formic (methanoic) acid – generated by wood, wood composite products such as medium-density fibreboard and by some adhesives and coatings, can also react with inorganic materials, often to produce white crystalline deposits on their surface. The formation of calcium acetates can affect the appearance of porous ceramics, particularly when stored in damp conditions, and is responsible for the white efflorescence on the surface of shell specimens first noted in the nineteenth century and termed Byne's disease.

In addition to gaseous pollutants, certain particles can trigger chemical changes. For example, chloride ions are present in salt-containing aerosols in coastal environments (Saunders and Kirby 2004), the salt used to grit streets and shed skin cells that have salts on their surface (a significant component of indoor dust). Another of the mechanisms by which vermilion (Hg_2S) deteriorates and changes colour is through its reaction with chloride ions to produce corderoite ($Hg_3S_2Cl_2$) and calomel (Hg_2Cl_2) (Spring 2016). These grey compounds can sometimes have a lilac tint that, in combination with the black of metallic mercury (see above), contributes to the hue of degraded vermilion (see Figure 4.5).

In very damp environments, particularly in buildings that are not well isolated from ground water or precipitation, chloride and other ions can be mobilised within the structure and encounter colorants with which they may react. The transformation of the blue pigment azurite into one of the green basic copper chlorides is a good example (Figure 4.11). Instances have been noted of the formation of either paratacamite (Dei et al. 1998), or atacamite (Naumova, Pisareva and Nechiporenko 1990); both are forms of the basic copper chloride $Cu_2(OH)_3Cl$. Although copper chlorides have been used as pigments in their own right, the pattern of alteration seen in Figure 4.11 and elsewhere suggests an alteration due to damp local conditions.

4.1.3.5 Carbon dioxide

As we saw for some metals, atmospheric carbon dioxide can often play a role in chemical changes. The lightening of red lead is noted less frequently than its darkening by oxidation to lead(IV) oxide or conversion to black lead sulfide – both described earlier. This colour change, which has been described in European and Asian objects (Saunders and Kirby 2004), is due to the transformation of red Pb_3O_4 to basic lead carbonate (lead white, $2PbCO_3 \cdot Pb(OH)_2$) by the action of carbon dioxide and moisture, often accelerated by the presence of light. While red lead is often less vulnerable in easel paintings, where it is protected from moisture by the binding medium to some extent, instances of lightening have nevertheless been noted (Saunders, Spring and Higgitt 2002, Spring 2016).

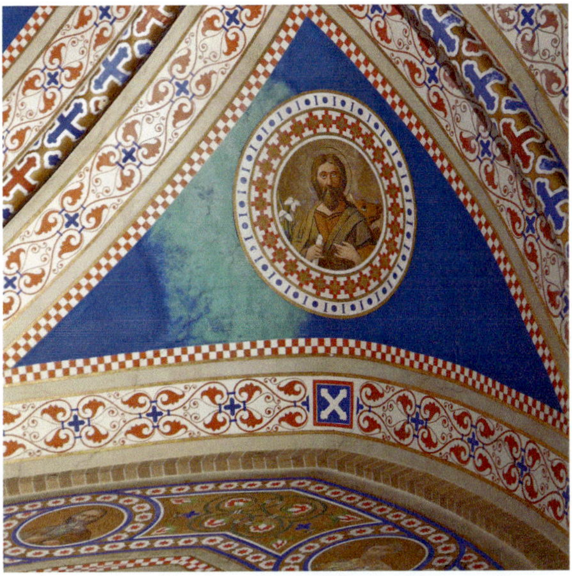

Figure 4.11 The ceiling of the Chapel of Saint Martin in Parma Cathedral. The azurite paint has discoloured where moisture has reached some parts of the vault. The green product is probably a basic copper chloride

4.1.3.6 Organic compounds

As we saw in Chapter 2, the chemistry of organic compounds is more complex than that of most inorganic materials. As a result, it is generally more difficult to pinpoint the changes to organic molecules that have led to an alteration in colour, particularly as the end products often remain unidentified and there are a number of ways in which the chromophore could be affected. Additionally, several competing chemical processes may be occurring, some of which are accelerated by light, heat or moisture. It is possible, however, to distinguish a few general reaction types that produce most of the changes that we might observe – and in so doing, identify those conditions that might contribute to alterations in colour.

As many organic colorants owe their colour to conjugated structures, a few of which are illustrated in Figures 2.18–2.20, reactions that lead to a loss of conjugation may shift the absorption band out of the visible region so that the molecule is no longer coloured. In contrast, we may encounter colour changes that result from increased conjugation and delocalisation in organic compounds, creating new chromophores; a common example of this phenomenon is the yellowing of colourless materials such as binders and coatings.

4.1.3.7 Oxidation

The oxidation of organic compounds can lead, successively, to the formation of alcohols, ketones or aldehydes and carboxylic acids, often also resulting in cleavage of the initial organic compound. The rates of reaction with atmospheric oxygen at ambient temperatures are generally relatively low, but are increased by heat or exposure to light or ultraviolet radiation. In addition, oxidation can be accelerated by certain pollutant gases that are strong oxidants, such as ozone or nitrogen dioxide, or by raised relative humidity (Modugno et al. 2019).

Heat, light and pollutants affect the rate of oxidation by increasing the number of excited species present. For example, radiation absorbed by molecules containing carbon-to-carbon (C=C) or carbon-to-oxygen double bonds (carbonyl groups, C=O) can produce a free radical species or an excited state (see *Radicals and excited states* box), either of which can act as precursors to a variety of further reactions.

Radicals and excited states (see Book 1 for more detail)

Radicals (sometimes termed free radicals) are atoms, molecules or ions that have at least one unpaired valence electron – that is in the outermost occupied shell in Figure 2.1. The unpaired electron makes the radical highly reactive. Relevant examples of radicals include the hydroxyl radical (HO·) and organic peroxyl radicals ($RO_2\cdot$).

Following convention, a free radical is indicated by the '·' symbol, and the letters R and R′ are used to represent all or part of an organic molecule.

Excited species are atoms, molecules or ions that, through the absorption of energy, exist in an excited electronic state – see Figure 4.2 – but do not contain unpaired valence electrons.

A relevant example of such an excited species is singlet oxygen (written as $^1O_2$).

In the example of light-induced-oxidation (photo-oxidation), a free radical produced from the organic compound by the action of light can react with atmospheric oxygen or another oxygen-containing molecule present in the material (Equation 4.2). Equally, the action of radiation can produce excited states of oxygen or oxygen radicals (see *Radicals and excited states* box) that may react with the organic compound (Equation 4.3).

$$RH \xrightarrow{h\nu} R\cdot + H\cdot$$

$$R\cdot + O_2 \longrightarrow RO + O\cdot$$

$$R\cdot + R'O \longrightarrow RO + R'\cdot \qquad \text{(Equation 4.2)}$$

A radical R· reacts with oxygen (O_2) or an oxygen-containing molecule (R'O)

$$O_2 \xrightarrow{h\nu} 2O\cdot$$

$$O\cdot + R \longrightarrow RO\cdot \quad \text{(Equation 4.3)}$$

An oxygen radical reacts with molecule R

In each reaction above, a new free radical is created. This can go on to react with other molecules, at which point additional energy, for example, from continued light exposure, may no longer be needed for further reactions to proceed.

Colour can be lost if the initial oxidation is followed by a fragmentation of an organic compound into smaller, less highly delocalised, molecules. This is frequently the case when large polymeric compounds break down into shorter-chain molecules. Additionally, when C=C bonds in an organic compound are oxidised to an alcohol, there is a potential reduction of conjugation and consequent loss of colour. Analogous oxidation reactions can also affect other chromophores, such as the –N=N– groups in azo dyes.

In contrast, further oxidation to ketones, aldehydes and carboxylic acids creates new C=O bonds that may increase conjugation, leading to the formation of new coloured species. Some natural and synthetic polymers are good examples of the latter; for example, polyethylene, polypropylene and polyvinyl chloride (PVC) yellow by radiation-induced oxidation that creates new carbonyl groups. Another common example of discoloration that results from the creation of new C=O bonds is the yellowing of paper with a high lignin content, where discoloration results from the oxidation of phenolic groups in lignin to quinones, which can be strongly coloured, see Figure 5.6. The formation of quinones is also implicated in the yellowing of aromatic polyurethanes.

Many traditional varnishes on western European easel paintings contain the triterpenoid resins mastic and dammar. Both yellow on ageing, primarily due to oxidation, which proceeds in the presence or absence of light (Dietemann et al. 2009). As varnishes containing these resins continue to be popular because they have useful optical properties, research has been conducted into the use of inhibitors that reduce their degradation and yellowing (de la Rie and McGlinchey 1990).

More complex reactions may also occur; for example, the initial products of photo-oxidation may combine to generate larger molecules or cross-links may form between natural or synthetic polymer chains. While these processes are important in film formation and the curing of plastics, they can also lead to issues such as increased brittleness and discoloration (see also Section 4.1.3.9).

Because photo-oxidation is a frequent cause of colour change and more general deterioration in organic materials, it is tempting to limit both the exposure to light and availability of oxygen. The former is possible and indeed commonly done, but placing individual objects in low-oxygen environments for long periods is difficult in practice. While mass storage of paper-based collections in low oxygen repositories is not uncommon, and rubber and vulnerable synthetic polymers have been stored in oxygen-free enclosures to good effect, it has proved difficult to display objects while excluding oxygen, see Section 7.6.

4.1.3.8 Hydrolysis

Many organic compounds react with water, or with the hydroxyl ions (OH^-) and hydroxyl radicals (OH·) generated from it, in a process termed hydrolysis. These reactions can proceed in the absence of radiation but are generally accelerated when its absorption generates excited species or free radicals that go on to react with moisture (Equation 4.4) or provides the energy to create hydroxyl radicals (Equation 4.5).

$$RH \xrightarrow{h\nu} R\cdot + H\cdot$$

$$R\cdot + H_2O \longrightarrow ROH + H\cdot \qquad \text{(Equation 4.4)}$$

A radical R· reacts with water (H_2O)

$$H_2O \xrightarrow{h\nu} HO\cdot + H\cdot$$

$$HO\cdot + RH \longrightarrow ROH + H\cdot \qquad \text{(Equation 4.5)}$$

A hydroxyl radical (OH·) reacts with molecule RH

Hydrolysis can lead to the formation of an alcohol where once there was a double bond, which may lead to a loss of conjugation that causes a compound to lose its colour.

Hydrolysis, particularly photohydrolysis, plays a role in the deterioration of a range of materials, such as the breakdown of cellulose molecules in paper. In common with oxidation, however, many studies of the effect of hydrolysis on cultural heritage objects have focused on its influence on their colour. A series of studies on the action of light on organic colorants, dating back to experiments by Russell and Abney in the late nineteenth century (Russell and Abney 1888), has found that the increased availability of moisture accelerates colour change. Research has also shown that if these colorants are used in a binding medium that repels moisture (a hydrophobic binder, such as a drying oil), they are protected from photohydrolysis to some extent, confirming the role of water in the process (Bailie, Johnston-Feller and Feller 1988; Saunders and Kirby 2001; 2004).

In acidic or alkaline conditions, hydrolysis can be instigated by hydrogen or hydroxyl ions respectively. Acidic conditions (low pH) may arise from the oxidation of organic molecules as described above or from acidic pollutants. Acid hydrolysis is one of the processes responsible for the breakdown of long chain natural polymers such as collagen in leather or cellulose in wood, textiles, canvas and paper. As the products of these reactions are themselves acidic, they can promote further degradation of the collagen or cellulose molecules, a process known as autocatalysis. To counter acid hydrolysis, an alkaline material is sometimes applied to the rear of paper or canvas substrates to reduce its acidity. At high pH, hydroxyl ions are the agent of hydrolysis; this situation is encountered less frequently, although some conservation treatments involve alkaline conditions.

In common with photohydrolysis, acid or alkaline hydrolysis may not primarily affect the colour of organic materials (in acid hydrolysis it is the mechanical strength of the paper that is compromised), but colour change may be the visible sign of these deteriorative processes.

4.1.3.9 Cross-linking

We have seen that when organic compounds absorb energy, particularly radiation, radicals may be formed that react with oxygen or water in their surroundings. Reactions with other organic molecules or radicals may also occur, one of the more significant examples of which is that between the long chain natural or synthetic polymers that form the substrates in many cultural heritage objects, a process termed cross-linking (Equation 4.6):

$$RH \xrightarrow{h\nu} R\cdot + H\cdot$$

$$R'H \xrightarrow{h\nu} R'\cdot + H\cdot$$

$$R\cdot + R'\cdot \longrightarrow R - R' \qquad \text{(Equation 4.6)}$$

Two radicals R· and R'· react with each other

Bonds between polymer chains can be beneficial, improving the strength of the material, but excessive cross-linking may make it too rigid, increasing its brittleness, or decrease its solubility. Alongside these changes to material properties, there may be more visible changes, for example, in the colour or transparency of the object.

4.1.3.10 The effect of pollutants

For gaseous pollutants there are three main routes leading to colour change. First, many pollutants are oxidants and the oxidative effects of ozone, nitrogen dioxide and peroxyacyl nitrates (PANs, which are found in photochemical smog) on organic colorants have been studied amid concerns about their possible effects on museum objects (Whitmore and Cass 1989).

Second, as we have already seen, in the presence of moisture, oxides of sulfur and nitrogen form acid solutions (containing sulfurous, sulfuric, nitrous or nitric acid depending on the nature of the pollutant gases). These acidic conditions accelerate the hydrolysis of many organic compounds, such as paper, textiles and many colorants (Williams, Grosjean and Grosjean 1993).

Third, direct reaction between pollutant gases and organic compounds may result in the formation of new chromophores. For example, NO_2 can cause the formation of compounds containing nitro (–NO_2) or nitroso (–ONO_2) groups, both of which owe their yellow colour to the –NO_2 chromophore.

It is often stated that dust particles, particularly those containing metals or some metal compounds, can act as centres for deteriorative reactions, but evidence for this, and examples of the damage that might result, are scant.

4.1.4 Inherent vice

Finally, we might consider two types of interaction that fall in a grey area between intrinsic and extrinsic reactions. Each is technically an extrinsic reaction and can be considered intrinsic only in the sense that the reactions involve materials present within, or generated by, the object rather than from the external environment. The effects of such processes are sometimes termed 'inherent vice', as the cause of deterioration lies within the object itself, making it potentially more difficult to address.

4.1.4.1 Pigment–binder interactions

Over a number of years, it has become apparent that the model of paints as an inert pigment held in place by a binder is over-simplistic. We know that binders undergo chemical reactions, as this affects how they dry after application, and in the preceding sections we have seen that many pigments are prone to chemical change. The pigment and binder may also react with each other, a situation that turns out to be far from uncommon across a range of painted objects and which now constitutes a burgeoning field of study, only briefly summarised here (see van Loon et al. 2021; Izzo et al. 2021).

Many such reactions are between metal-containing pigments (or additives added to help the paint dry) and the free fatty acids present in drying oils and some other binding media; the products of these reactions are termed metal soaps. Although the external environment or past conservation treatments (especially those involving moisture) may well play a role, particularly influencing the rate of reaction, these interactions generally also proceed in the absence of such factors.

Because pigments based on lead are quite common (lead white, red lead, etc.) and lead-containing materials have frequently been used as driers for oil paint, lead soaps are often encountered in paint that is decades-old and very commonly in paint that is centuries-old. As we will see in Chapter 5, their formation can affect the appearance of paint films in several ways. First, pigments including red lead and massicot will lose their colour, since lead soaps are typically colourless, as are the secondary products that might form subsequently. Second, as lead soaps are generally more transparent than the pigments from which they form, the overall transparency of the paint layer may increase, revealing underlying colours. Third, migration of lead soaps to the surface of a paint film can produce a white veiling film or, where the soaps aggregate, a speckled appearance that is particularly noticeable in dark paint.

Lead-containing materials are by no means the only source of such changes and there is much current interest in the formation of metal soaps by zinc-containing pigments and paint additives. It is, however, worth noting that colour change is only one of the outcomes of metal soap formation, which may pose a more serious threat to the integrity of the paint layers.

Reactions between pigment and binder do not necessarily produce a colour change in the pigment component of the system. Some metal ions present in pigments initiate or catalyse those reactions in the binding medium that we encountered above, increasing conjugation or cross-linking and leading to its yellowing or darkening. We will look at two examples – for copper- and cobalt-containing compounds – in Chapter 5.

4.1.4.2 Intrinsic pollutants and autocatalysis

We have seen that pollutant gases from atmospheric sources or generated in display environments can affect the colour of materials. However, in some objects the source of deterioration comes from within, either because a component in a composite object generates a substance that causes deterioration in another material, or because once deterioration has begun, its by-products catalyse further change through autocatalysis.

In composite objects, natural or synthetic polymers are frequently the source of intrinsic pollutants, as their breakdown may liberate species that react with other parts of the object. Wool and silk are sources of reduced sulfur gases that, as we have seen, may discolour metals, especially silver, and certain pigments. This can be an issue for costume or armour collections and is particularly relevant to textiles containing metal threads, which often comprise strips of silver leaf wound around a silk core, see Figure 4.7. In instances where the sulfur responsible for darkening colorants such as red lead, lead white or copper-containing pigments might derive from sulfur-containing pigments (e.g., orpiment and realgar), artists' manuals frequently warned against using these incompatible materials in mixtures or in close proximity.

Oils occurring naturally in objects or applied to them to improve appearance or suppleness may also interact with reactive metals; for example, blue–green copper stearate has been found on brass studs in leather objects.

We saw earlier that wood liberates acetic acid vapour, as does the hydrolysis of cellulose acetate. Acetic acid may affect metal components in the object, particularly those made from lead. The ready degradation of another early synthetic polymer, cellulose nitrate, liberates nitrogen oxides, which may hydrolyse to produce nitric acid, a species whose effects we have already seen. Many adhesives, consolidants

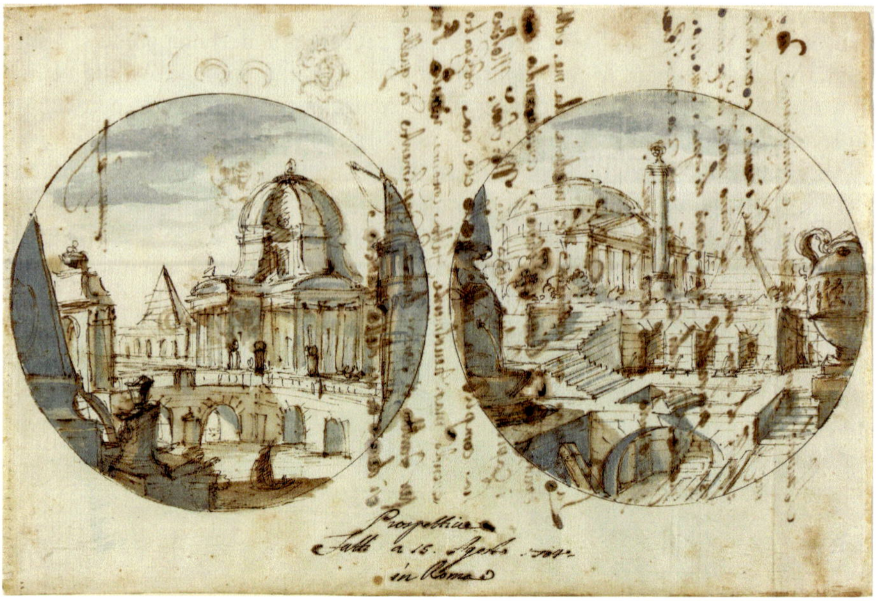

Figure 4.12 A pen and ink drawing depicting *Architectural Fantasies with Temples and Pyramids*. The iron gall ink used in this drawing and the text on the verso has reacted with the paper, causing browning and small damages to the structure of the paper. [Source: Filippo Juvarra, 1704. 20.3 × 29.9 cm. National Gallery of Art, Washington, DC, 1989.6.1. Ailsa Mellon Bruce Fund.]

and coatings applied during previous conservation treatments – including cellulose acetate and cellulose nitrate – may also contribute to deterioration. As the hydrolysis of cellulose acetate and depolymerisation of cellulose nitrate are both autocatalytic, discoloration and deterioration of these polymers, and their ill effects on other components, can proceed very rapidly.

Previously, we saw that autocatalysis contributes to the acid hydrolysis of paper. A further example of its effects is offered by another phenomenon affecting paper-based collections. Black inks and dyes based on iron and tannin have been used on a wide variety of organic substates by cultures worldwide. The best-studied examples are those of iron gall inks, traditionally prepared from an iron salt and tannin from oak galls. In drawings and manuscripts, compounds from the ink catalyse oxidation and acid hydrolysis that first discolour the substrate and can ultimately lead to its disintegration (Figure 4.12) (Kolar and Strlič 2006).

4.2 Physical changes

The sections above explored how chemical changes affect those materials that owe their colour to their chemistry (chemical colours). Here, we will consider physical changes that affect either physical or chemical colours before finally examining those chemical reactions that lead to a change in the structure of physical colours, affecting their appearance.

4.2.1 Physical effects from physical changes

4.2.1.1 Physical forces

Physical forces can alter structure of an object, particularly its surface properties, modifying the way in which it interacts with light. We saw in Chapter 2 that interference colours, particularly those associated with insects' wings or birds' feathers, arise from their fine structure; if this is crushed or distorted – perhaps through inappropriate handling or storage – their colour can be changed or lost.

While the intrinsic colour of an object may be unchanged, flattening or crumpling of its surface by physical forces can change its appearance markedly by altering the way that it interacts with light, for example, the flattening of a textile or impastoed painting.

The dimensional changes that may result from fluctuations in temperature or from swelling and shrinkage in hygroscopic materials (particularly cyclical swelling and shrinkage) can lead to alterations in the physical structure of objects that affect their interaction with light. A common example is the development of networks of fine cracks on wooden, ceramic and painted surfaces, variously termed crazing or craquelure depending on context. These cracks increase scatter at the surface, leading to a desaturation in colour and loss of gloss. The stresses that build up in an object as its dimensions change may also be relieved by the separation of materials, creating internal voids and delaminations that increase scatter within the object; again, these typically desaturate the colour and increase its lightness.

The scattering properties of materials can also be affected by the scratching or scouring of their surface, which may result from inappropriate handling and storage, or from surface abrasion by dust. The increased roughness that results will tend to reduce surface gloss and to desaturate and lighten underlying colours.

As conservators, we need to be careful not to expose objects to excessive physical forces during treatment – and we are now better at recognising when this might be an issue. For example, dust removal should avoid dragging material across delicate surfaces, and in treatments that require pressure – such as lining paintings – we consider the effect on texture.

4.2.1.2 Deposition and movement

While the air in many museums is now cleaner than in the past, the deposition of dirt on objects on open display is a familiar issue for most conservators. Although the composition of dirt will depend on factors not covered here, the overall result will tend to be the obscuration of underlying colours and a reduction in gloss. Depending on context and display conditions, other materials that may accrete on objects include candle wax and grease from visitors' hands. The latter will tend to saturate the colour of porous materials such as stone (sculptures seem frequently to attract touching) but may also create a sticky surface that retains a higher proportion of dust.

Fortunately, both dust and grease can be removed from many types of surfaces. The ease with which dust can be removed depends on its stickiness and the extent to which it has penetrated the layers beneath. As mentioned above, even when the dust layer is superficial, care is needed to avoid abrading the object. Grease removal from robust surfaces typically employs solvents, poultices and in some cases laser cleaning.

In addition to the deposition of material on the surfaces of objects, there are processes that transport substances within or through them. Many involve water and the dissolution and reprecipitation of materials in a new location. In objects that are connected to buildings, such as wall paintings and sculptures, water transport of soluble salts from that structure, or the ground, may produce (often white) crystalline deposits on their surfaces. In some cases, this veiling layer desaturates or obscures the colour beneath, but at worst the dimensional changes caused by the formation of such solids can result in losses from the painted surface. Similar phenomena may be seen in freestanding objects, including porous ceramics, that have been stored in very damp conditions or involved in floods. Other materials that may migrate and crystallise on the surface include fats, which may derive from the original object, from materials used in former treatments or from their traditional community use, particularly in the case of sacred objects.

Colour changes can also result from the solubilisation of dyes and inks in, for example, textiles and works of art on paper. This irreversible 'bleeding' of water-soluble colorants into white or light-coloured substrates may be triggered by very damp conditions, inundation, or inappropriate aqueous conservation treatments.

4.2.2 Physical effects from chemical changes

We have seen that chemical reactions involving colorant molecules can lead to the loss of a chromophore. However, chemical reactions involving other molecules can also be responsible for changes to the structure of objects that affect the way they interact with light, and thus their colour or appearance.

We saw above that structural colours, such as those found in feathers and insect wings, can be lost or altered by changes in their geometry that alter their interaction with light. Keratin, the protein found in birds' feathers, and chitin, the polymer in insects' wings, are relatively stable materials, although both can be degraded by reactions that are triggered by exposure to ultraviolet radiation. Such reactions may cause distortion or collapse of the structures responsible for interference colours in both biological specimens and other collections, for example, costume collections in which feathers are present.

As we will see in more detail in Section 5.5, which describes changes in transparency, some chemical processes cause changes in the refractive index of a material. This is sometimes the case as a binding medium ages, and – if its refractive index increases and becomes closer to that of the pigment particles contained within the

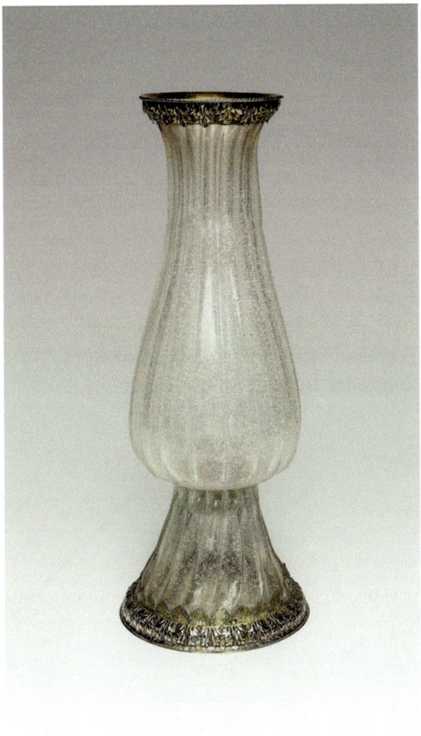

Figure 4.13 One of a pair of glass vases. The inherently unstable glass shows a fine network of cracks (crizzling) that produces a hazy appearance. [Source: Northern Netherlands, *c*.1675–1685. 42.9 × 15.9 cm. Art Institute of Chicago, 2016.439.2. Gift of Neville and John Bryan.]

layer – the transparency of the film will increase (see Figure 1.8). As described earlier, some pigment–binder reactions may lead to the formation of compounds, for example, metal soaps, that are more transparent than the original materials in the layer.

Chemical deterioration of the matrix of an object can lead to changes in its dimensions, creating voids and fissures, or lower the cohesion between materials, making them more vulnerable to delamination. We see examples of such behaviour in unstable glasses, where leaching of components of the matrix may lead initially to a fine network of cracks (termed crizzling) that alters the appearance of the object through greater light scattering at or near the surface (Figure 4.13), and eventually to structural failure of the object.

Many of the phenomena seen in paintings can also be explained – at least in part – by the changes to the optical properties of the paint and varnish films that result from ageing of these organic layers or their interaction with moisture. At the

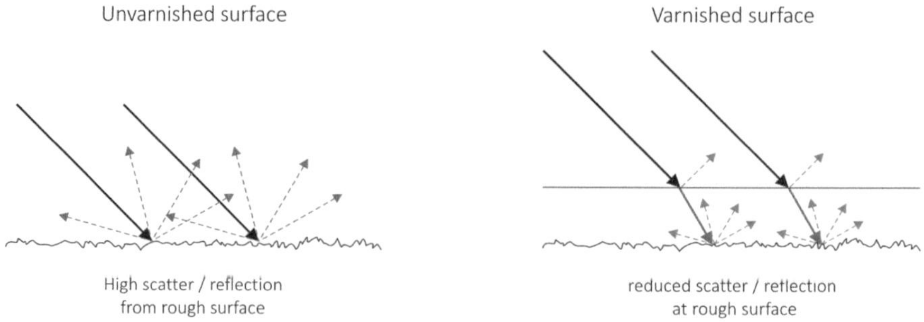

Figure 4.14 Schematic representation of the effect of applying a varnish or coating on reflection and scatter from a rough surface

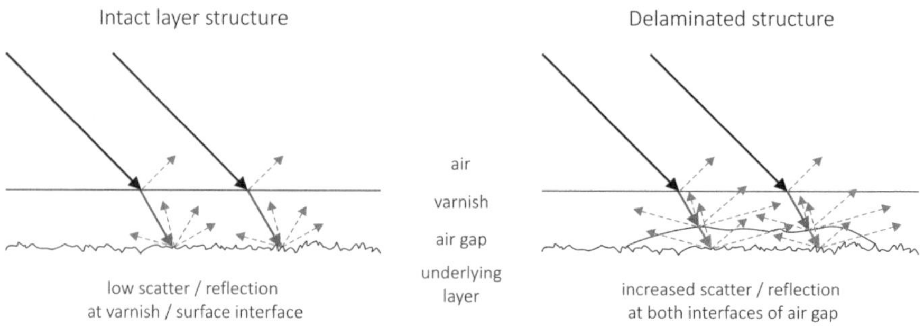

Figure 4.15 Schematic representation of the effect of delamination between layers on reflection and scatter at the interfaces between layers

surface, the deterioration of the binding medium or varnish can expose a rougher layer that is more scattering and less glossy (Figure 4.14), while delamination can lead to scatter between paint layers, or between paint and varnish layers, that gives a milky or blanched appearance (Figure 4.15).

Finally, chemical reactions within the object can lead to the formation of layers on the surface. These can vary from thin films that may impart a cloudy appearance – a phenomenon often called 'bloom' when it appears in coatings and varnishes – to thicker crusts that obscure the underlying colour. As we saw in Section 4.1.4.1, such crusts may arise from the migration of metal soaps to the object's surface.

5 How colours change

In Chapter 4, we explored the chemical and physical processes that give rise to changes in colour, but our day-to-day concerns about such alterations tend to focus on how the appearance of an object or area is changed, rather than the underlying mechanism. Changes in colour can, for example, alter the balance (harmony) of colours in an object, reduce our ability to see and distinguish different areas of a whole and disconnect the current appearance and its interpretation from that intended by the artist or creator.

5.1 Lightening (fading)

Lightening is perhaps the most common phenomenon that we encounter, with an increase in lightness usually accompanied by a decrease in the strength of a colour. In colorimetric terms we can define lightening as an increase in L* (our measure for lightness), alongside a decrease in chroma (C* or colourfulness). In the example of the lightening of eosin (also sometimes termed geranium lake) in Figure 5.1 the colour that is being lost is a red, so a plot of L* and a* in the lower part of the figure is a good means to chart change as the colour lightens and moves towards the neutral axis of the CIE Lab diagram. The upper part of Figure 5.1 shows a series of spectra that chart the general increased reflectivity across the visible spectrum and the loss of the characteristic peaks to which the original pigment owes its colour.

We often use the term fading to describe such changes, as on many – but not all – occasions it results from the gradual loss of a colorant. In these instances of fading, the increase in lightness may arise for two reasons.

First, where a chemical change in the colorant molecule destroys the chromophore to yield a white material. As we saw in Chapter 4, some metals and inorganic pigments behave in this way, for example, the patination of aluminium, the oxidation of yellow or orange arsenic sulfides to the white arsenic oxide and the conversion of red lead to white lead carbonate.

A second, more common, reason for fading is that the loss of colorant reveals a white or pale-coloured material in a substrate or mixture. In a simple example, the colour in a thin watercolour wash on a white paper may become dominated by the increasing visibility of the white substrate as the colorant reacts to form colourless

DOI: 10.4324/9781003397182-6

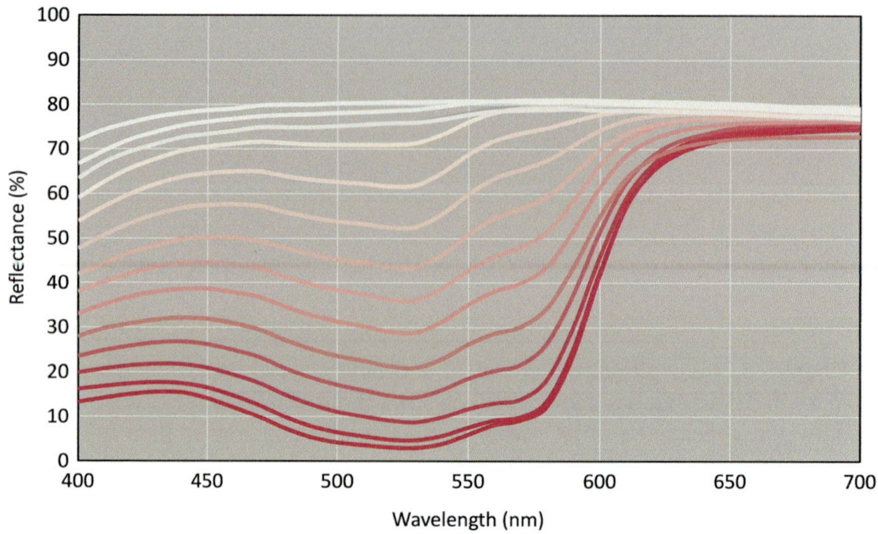

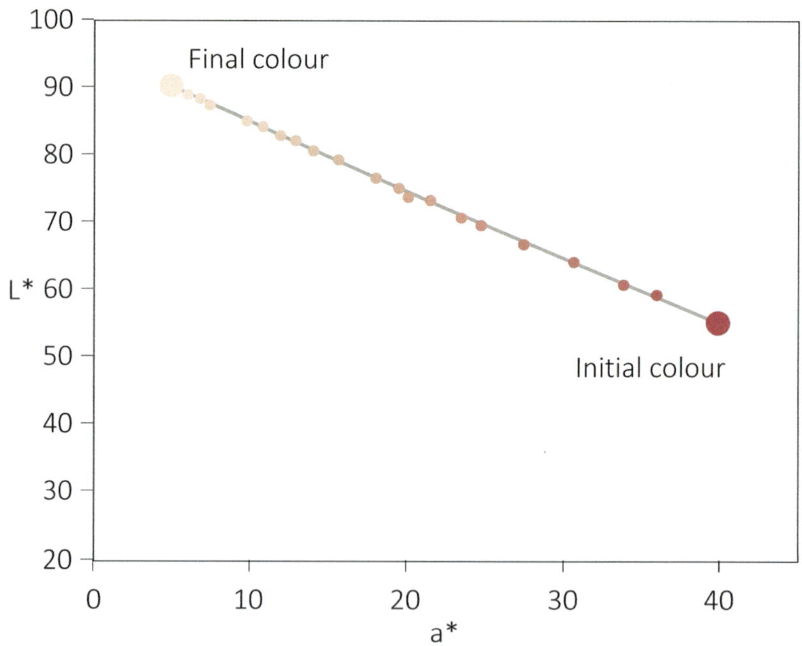

Figure 5.1 The lightening of eosin (geranium lake). The upper part of the figure shows reflectance spectra recorded as the colorant changes. The overall reflectivity increases and the characteristic absorbances that give rise to the red colour are lost. The lower part of the figure shows the same changes colorimetrically. As the lightness (L*) increases, redness (a*) decreases

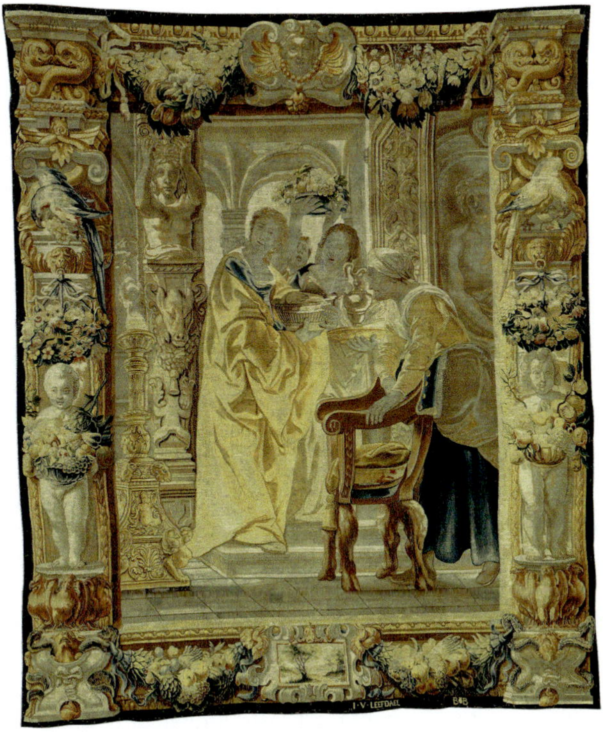

Figure 5.2 Flemish tapestry depicting *Four Servants*. The limited range of colours is probably due to the fading of natural red and yellow dyestuffs. [Source: Design by Jacob Jordaens, woven at the workshop of Jan van Leefdael, *c*.1650. 351.2 × 424.8 cm. Art Institute of Chicago, 1890.8. Gift of Charles L. Hutchinson through the Antiquarian Society.]

or less strongly coloured compounds (Figure 4.3). In other cases, the loss of colorant reveals the white or pale-coloured threads in a dyed fabric (Figure 5.2), a white or off-white preparatory layer in a painting or the white substrate onto which a dyestuff has been precipitated when making a lake pigment.

Fading of this type is more common in organic molecules in which, as we saw in Chapter 4, the loss of the chromophore is frequently caused either by fragmentation of the molecule or by a loss of conjugation that shifts the absorption band out of the visible region so that the molecule is no longer coloured. Many natural dyestuffs, such as those derived from plant and insect sources that are illustrated in Figures 2.18–2.20, and the lake pigments derived from some of them, are prone to fade, particularly when exposed to light (Saunders and Kirby 1994).

Despite their tendency to fade, these natural materials were long an essential part of the artist's or dyer's palette. They have largely been superseded by more stable synthetic organic colorants, although there are some well-reported examples of modern colorants that are prone to fading, particularly when exposed to light. The fading of both eosin (geranium lake) and lithol red have been studied extensively, perhaps because of their use by well-known artists – Vincent Van Gogh (Townsend and Freemantle 2023) and Mark Rothko (Standeven 2008) respectively.

Not all instances of fading are the result of the loss of a colorant that has been added deliberately. For example, the bleaching of many materials, including wood, ivory, textiles and paper may result from the loss on ageing or light exposure of coloured molecules that occur naturally in the plant or animal sources from which they are obtained.

Fading can severely affect the appearance and interpretation of objects and often has the greatest impact where the initial dye or pigment concentration was low, for example, in thin watercolour washes. Even when more colorant is used, severe fading can alter the relative strength of colours. In the colour samples shown in Figure 5.3 the red lake pigment survives only where its concentration was initially very high, with the mid-tones and highlights essentially indistinguishable after light exposure. In paintings, only the deepest shadows are retained, while any modelling is lost as the mid-tones become as pale as the lightest hues. This behaviour has been attributed partly to the lower concentration of red colorant in the mixtures with white, but also to inter-reflection of light within the paint layer by the white pigment, which can exacerbate fading of lighter tones (Johnston-Feller et al. 1984).

In some cases, our perception of objects is shaped by widespread change; we are, for example, so used to seeing tapestries dominated by the more stable blues and the neutral yarn colours left when yellow and red natural dyestuffs fade that this is how we may expect them to appear (see Figure 5.2).

An extreme example of what has been termed fading is offered by certain drawings in silverpoint executed by Leonardo da Vinci. Now invisible, they are only revealed when illuminated with ultraviolet radiation, which causes the paper

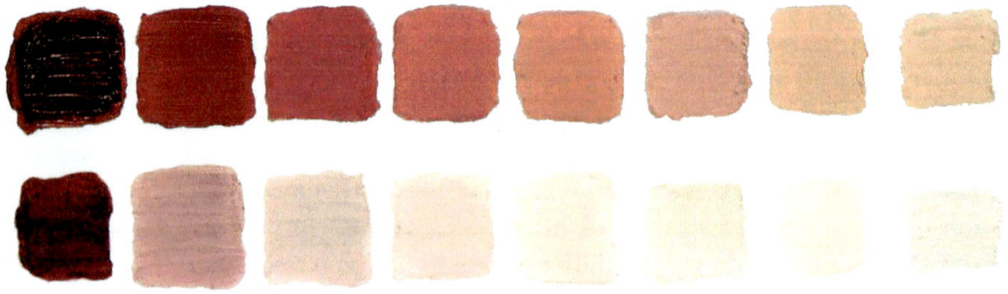

Figure 5.3 Pure madder lake pigment and mixtures with different proportions of lead white. From left to right: pure, 1:1, 1:4, 1:10, 1:20, 1:50, 1:100 and 1:200. The upper series has been kept in the dark, while the lower series has been exposed to light, causing severe fading that has rendered the mid and light tones virtually indistinguishable

(but not the drawing) to fluoresce. Analysis suggests that the change in appearance results from a loss of material from the surface rather than a chemical reaction peculiar to the silverpoint used in these drawings (Russell and Donnithorne 2014, esp. Figure 1).

Having examined changes that occur through the loss of colorant molecules, we need also to consider lightening caused by the obscuration of colours by a change in the surface that causes more light to be scattered or by a layer of white material. As we saw in Chapter 4, such scattering or white layers may result from physical or chemical changes at the surface of an object.

Lightening can arise from increased surface scattering caused by the development of fine craquelure patterns (see Figure 4.13). This not only increases lightness but, because this scattered light has not had the opportunity to interact with the colorants beneath, desaturates the colour. Surface scatter can also increase where a once smooth surface has become more textured due to abrasion or because a coating or varnish 'sinks'; again, the resultant rise in scattering increases the lightness and desaturates the colour.

Within the object, delamination or a loss of cohesion between binding medium and pigment particles will have a similar effect, lightening its appearance. For example, if delamination occurs between a varnish and an underlying layer, scattering will occur at both the varnish–air and air–substrate interfaces that have been created (see Figure 4.15).

As we saw in Chapter 4, chemical changes can lead to the creation of white material at the surface of an object that can veil or obscure the colour beneath, for example, the white patina on certain metals. Such a layer can also sometimes be caused by a once-transparent coating, such as a varnish, reacting with moisture or with components of the layers beneath. Alternatively, it may result from the migration of materials to the surface – either those created by reactions within, for example, a paint layer or glass vessel, or transported through the object by moisture, as might be the case for wall paintings and porous ceramics, where salts – many of them white – can recrystallise on the surface.

The phenomenon of a white veiling layer in easel paintings – sometimes termed blanching – has been studied extensively and does not seem to be attributable to a single cause. The deterioration of certain pigments, including yellow lakes, green earths and ultramarine has been implicated, as have discontinuities between the pigment particles and binding medium (mentioned above) that might lead to increased scatter. Recent research has found deposits of metal soaps, oxalates and sulfates at the surface of affected paint layers (as discussed in Section 4.1.4.1). These materials, created by reactions within the paint layers or through deterioration of the binding medium can form either a continuous thin veiling layer or protrude through the outer layer, increasing the surface roughness and light scattering.

In some circumstances, veiling layers can be caused by biological activity, as mould develops on the surface of objects displayed or stored in inappropriate conditions.

5.2 Darkening

Although less common than lightening or fading, there are many instances in which objects show darkening, that is, a reduction of the lightness of a material, which – as we saw in Chapter 1 – is accompanied by reduced colourfulness through the Hunt effect. In colorimetric terms we can define darkening as a decrease in L* accompanied by a decrease in chroma (C* or colourfulness). As an example, the darkening of vermilion can be expressed through a plot of L* and a*, the lower part of Figure 5.4, which shows the colour darkening and moving towards the neutral axis of the CIE Lab diagram as redness (a*) and colourfulness are lost. The series of spectra in the upper part of Figure 5.4 show the gradual reduction in overall reflectivity as vermilion darkens.

We might consider two routes to darkening in cultural heritage objects. First, chemical reactions that lead to the formation of black or dark-coloured materials. In Chapter 4 we saw that some metals, including silver, are prone to darken. Greater darkening is observed when objects are exposed to atmospheric pollutants that cause the formation of metal sulfides and it is notable that areas of silver leaf in manuscripts that are customarily kept closed tend to be better preserved than those in objects on open display.

Changes to some red and yellow pigments – including red lead, vermilion, massicot, chrome yellow and cadmium-based colours, as discussed in Chapter 4 – can lead to darkening. In alkaline environments, for example, in wall paintings (Gutscher et al. 1989), azurite has been observed to darken from blue to black due to the formation of tenorite, copper(II) oxide. However, the most dramatic alteration to an inorganic pigment is that of lead white, whose reactions with oxygen or reduced sulfide gases, which we encountered earlier, produce brown PbO_2 or black PbS respectively. These changes can be particularly disturbing because they can reverse the tonal range of an image, converting white highlights into areas of deep shadow (see Figure 4.9). See also Section 5.9.

Although organic pigments tend not to darken, initially colourless substrates, binding media and coatings can gain colour as they age. This often manifests as a yellowing of these materials, which is covered in Section 5.3. However, under some circumstances, this yellowing can give rise to a more general darkening, particularly when the optical effect of a changed binding medium combines with that of the pigment particles it contains through subtractive mixing. Examples of such changes can be found in paints based on copper-containing blues or greens; the pigment may promote reactions in the binding medium that lead to its discoloration to a yellow or orange hue. As these colours are complementary to those of the blue or green pigment, these areas will appear brown or black to the viewer.

A second cause of darkening is the accumulation of dirt or other dark-coloured material on an object. Some particles derive from the breakdown of sand, soil, or building materials, such as stone, concrete, brick and plaster, while others are generated by burning fossil fuels, and are loosely termed soot. Finally, skin, hair and textile fibres shed by visitors are major components of indoor dust.

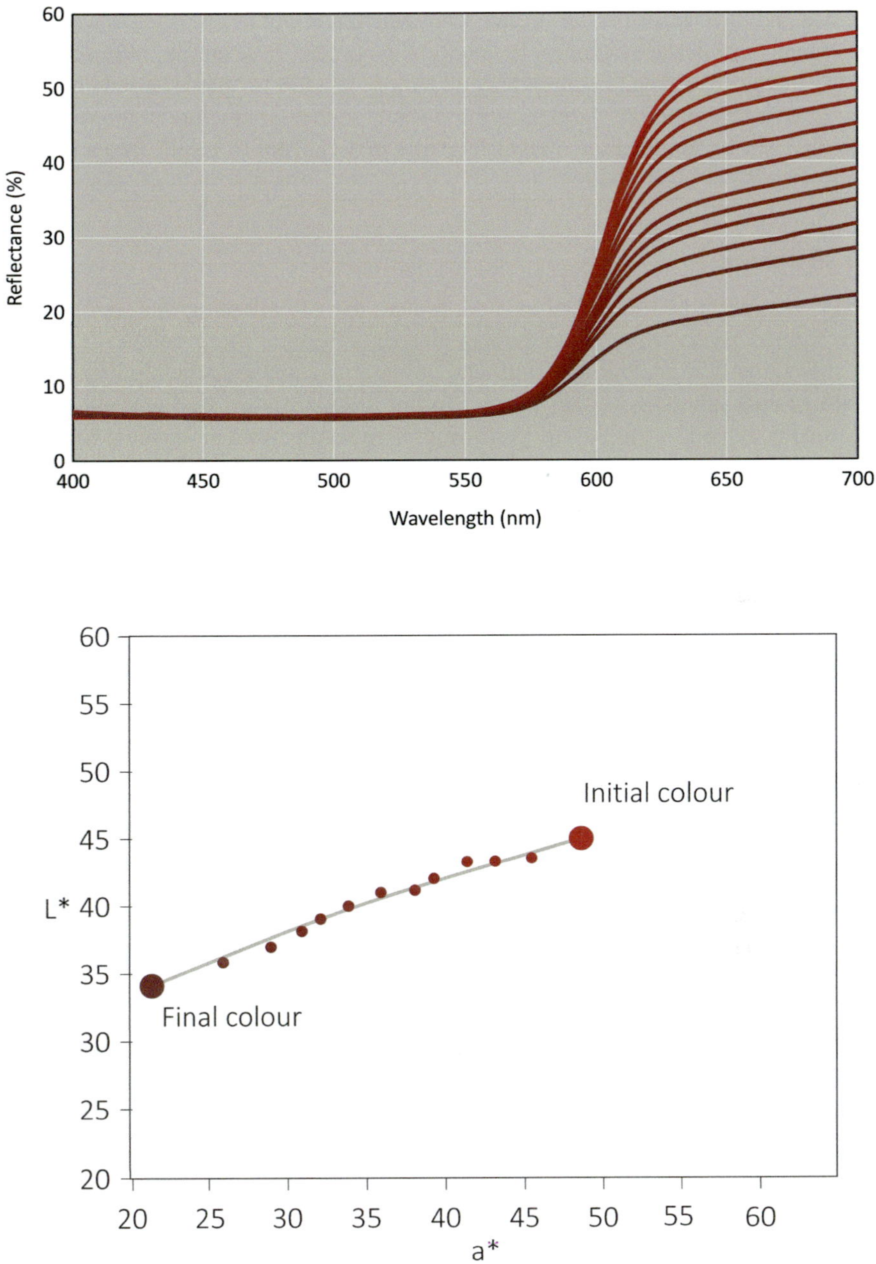

Figure 5.4 The darkening of vermilion. The upper part of the figure shows reflectance spectra recorded as the pigment changes. Although the characteristic reflectance in the red region of the spectrum is retained, the overall reflectivity decreases. The lower part of the figure shows the same changes colorimetrically. The lightness (L*) and redness (a*) both decrease as the colour darkens

The extent to which such darkening is reversible depends on the nature of the deposit and the fragility of the surface from which it might be removed, see Section 7.5. Some particles are rather more 'sticky' than others, for example, skin cells and tarry soot particles from incomplete combustion. In addition, particles may become ingrained into a material – in gaps in the surface, such as craquelure or the weave of a textile – or be imbibed into the surface of a material, for example, into a plastic, paint, varnish or other coating.

5.3 Yellowing

Yellowing might be thought of as a particular case of darkening; the term is usually applied when white or colourless materials become more yellow. It occurs most frequently in natural and synthetic polymers, resins, organic binding media, coatings and adhesives (either present in the original object or from subsequent treatment). In contrast to fading, yellowing often results from an increase in conjugation (perhaps through the formation of more double bonds or by cross-linking) that brings the absorption of molecules into the visible region.

Although the degree to which a colour has yellowed can be expressed through the yellowness index introduced in Section 3.8, we can also present yellowing colorimetrically on a CIE Lab diagram. It will be most useful to look at the b* axis, positive values of which define 'yellowness' (Figure 5.5). In the lower part of the figure it is plotted against L* and we can see that the lightness does not change greatly, as many strong yellows have a high lightness. In the upper part of Figure 5.5, we can also see how the reflectance spectrum changes as the sample yellows, with the absorption edge moving further into the blue end of the visible spectrum, so that the reflected radiation becomes richer in the red, orange and yellow wavelengths.

As we saw in Chapter 4, both natural and synthetic polymers have a tendency to yellow, often through photo-oxidation. Cellulosic materials, such as paper (particularly low grade paper: Figure 5.6), wood, canvas or cotton and proteinaceous polymers – including wool and silk – are ubiquitous in museum and other collections.

As plastics become an increasingly common component in collections, the susceptibility of those such as polyethylene, polypropylene, PVC and aromatic polyurethanes to yellowing (and darkening and loss of strength on prolonged exposure) is a concern for conservators. So-called high-performance plastics, including polyesters, polycarbonates and polyamides are less susceptible to change but additives, such as antioxidants, can also be a source of discoloration in the longer term, as the additives themselves degrade.

The binding media used in paints are prone to yellow on ageing. Particular attention has been paid to the yellowing of drying oils, not least to understand how empirical choices made by artists relate to the chemistry of the oils. Artists' manuals traditionally advised using certain oils, for example, walnut, with pigments whose appearance will most be affected by yellowing, such as whites or blues. The degree of yellowing on ageing is correlated to the extent of unsaturation in the long-chain

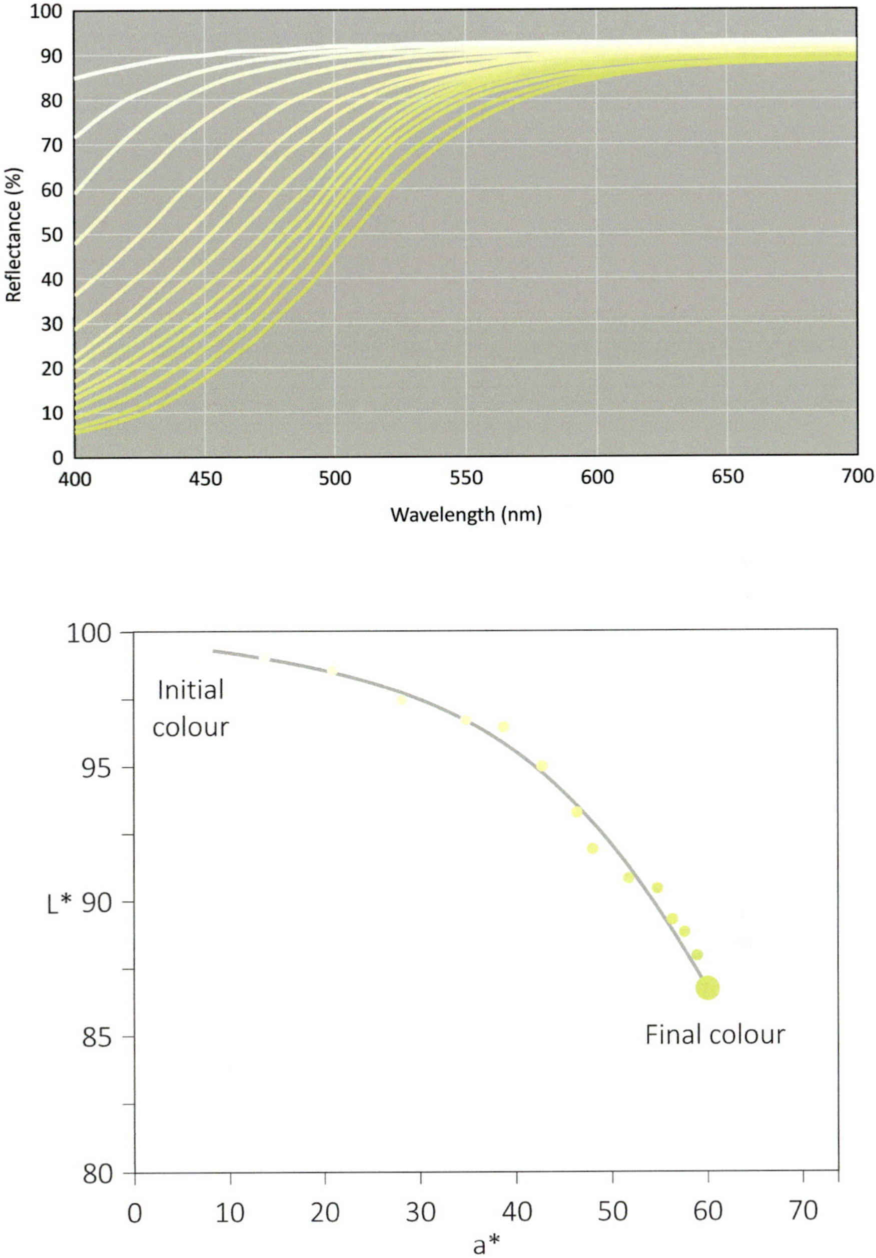

Figure 5.5 The yellowing of polyethene. The upper part of the figure shows reflectance spectra recorded as the white polymer discolours. The reflection in the blue region of the spectrum gradually decreases, while that in the yellow region remains high. The lower part of the figure shows the same changes colorimetrically. The yellowness (b*) increases appreciably, while the lightness (L*) only decreases slightly (as many yellows have high lightness)

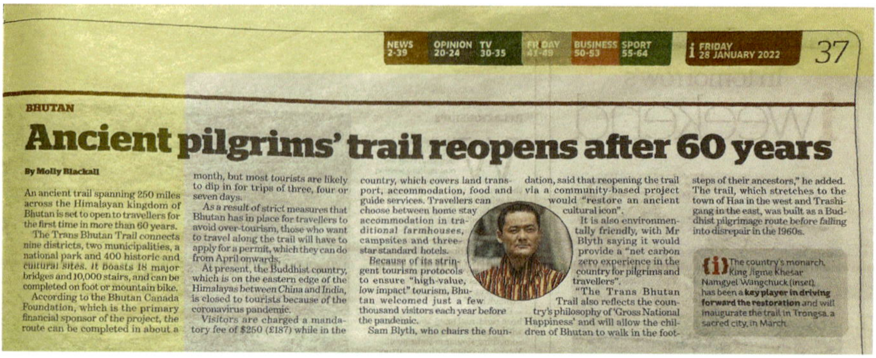

NEWS 2-39 OPINION 20-24 TV 30-35 BUSINESS 50-53 SPORT 55-64 FRIDAY 28 JANUARY 2022 37

BHUTAN

Ancient pilgrims' trail reopens after 60 years

By Molly Blackall

An ancient trail spanning 250 miles across the Himalayan kingdom of Bhutan is set to open to travellers for the first time in more than 60 years.

The Trans Bhutan Trail connects nine districts, two municipalities, a national park and 400 historic and cultural sites. It boasts 18 major bridges and 10,000 stairs, and can be completed on foot or mountain bike.

According to the Bhutan Canada Foundation, which is the primary financial sponsor of the project, the route can be completed in about a month, but most tourists are likely to dip in for trips of three, four or seven days.

As a result of strict measures that Bhutan has in place for travellers to avoid over-tourism, those who want to travel along the trail will have to apply for a permit, which they can do from April onwards.

At present, the Buddhist country, which is on the eastern edge of the Himalayas between China and India, is closed to tourists because of the coronavirus pandemic.

Visitors are charged a mandatory fee of $250 (£187) while in the country, which covers land transport, accommodation, food and guide services. Travellers can choose between home stay accommodation in traditional farmhouses, campsites and three-star standard hotels.

Because of its stringent tourism protocols to ensure "high-value, low impact" tourism, Bhutan welcomed just a few thousand visitors each year before the pandemic.

Sam Blyth, who chairs the foundation, said that reopening the trail via a community-based project would "restore an ancient cultural icon".

It is also environmentally friendly, with Mr Blyth saying it would provide a "net carbon zero experience in the country for pilgrims and travellers".

"The Trans Bhutan Trail also reflects the country's philosophy of 'Gross National Happiness' and will allow the children of Bhutan to walk in the footsteps of their ancestors," he added.

The trail, which stretches to the town of Haa in the west and Trashigang in the east, was built as a Buddhist pilgrimage route before falling into disrepair in the 1960s.

{i} The country's monarch, King Jigme Khesar Namgyel Wangchuck (inset), has been a **key player in driving forward the restoration** and will inaugurate the trail in Trongsa, a sacred city, in March.

Figure 5.6 Discoloration of low-quality newsprint after a few weeks' exposure to direct sunlight; part of the page that was protected from light has not yellowed

fatty acids in the oil, with walnut and poppyseed oils both having lower unsaturation than the more generally used linseed oil. Some unsaturation is, however, essential for drying, so while more saturated oils will yellow less, they may show issues connected with poor drying. Yellowing has also been found to be accelerated by certain metal-containing pigments, such as lead white and various copper- or cobalt-containing pigments.

When oil films dry in the dark, they can develop a yellow colour, again most noticeable in white or blue areas. The same is true of oil/alkyd paints. This initial discoloration bleaches quite rapidly when the paint is first exposed to light but will gradually redevelop on ageing. One consequence of this initial bleaching is that in studies of light ageing of materials, it is quite common to discard the first few hours of measurements, which may be dominated by this short-term effect. Similarly, it complicates the interpretation of microfade testing applied to such paints.

As mentioned earlier, varnishes containing natural resins also yellow as they age. As these are sacrificial coatings this might seem unimportant, as they can be removed in due course, but each removal carries a risk to the object, and the yellowed varnish will obscure the colours beneath while in place. Accordingly, considerable effort has been devoted to finding resins that yellow less and to the development of varnish additives that slow down this process in newly-applied natural resin varnishes (de la Rie and McGlinchey 1990).

Similarly, other conservation materials, such as adhesives, consolidants and coatings – often based on synthetic or natural polymers – are also prone to yellowing, e.g., through photo-oxidation. Extensive testing over many decades has identified those materials that are least prone to alteration (Down 2001).

5.4 Hue shifts

The changes we have considered so far involve either the lightening or darkening of a particular colour, or the development of colour by a white or colourless material in the specific case of yellowing (it is worth noting that there are other such changes, including the development of a purple colour in some manganese-containing glasses, thought to result from the creation of Mn^{3+} ions).

In other instances, we see a change that results from a shift in hue, as defined in Figure 3.12; so, for example, a green colour might become yellower or a purple bluer. Figure 5.7 gives spectral and colorimetric data for a typical hue shift, from green to blue in this case. Although we have plotted the a* and b* values, we can see that the hue angle changes from 53° for the original green to 303° for the final blue, and also how the chroma alters. In the spectra we see the shift of the reflection maximum from the green towards the blue portion of the visible region. British watercolours (usually landscapes) from the mid- and later eighteenth century, usually thinly painted, provide a good example of objects in which the type of hue shift illustrated in Figure 5.7 might occur. These often include mixed green tones made from a moderately lightfast blue and a more light-sensitive yellow pigment, many of which have a bluish cast today.

Some shifts in hue arise from a physical or chemical change to a material that alters its interaction with light. We have encountered several examples in Chapter 4, including green patinas on copper (see Figure 4.6), the alteration of orange realgar to yellow pararealgar, the conversion of blue azurite to green basic copper chloride (see Figure 4.11) and the development of a lilac sheen on some areas of vermilion (see Figure 4.5).

In Chapter 4, we also encountered the tendency of black iron–tannin colorants – which have been used as inks and to dye fabrics, barkcloth and basketry among other objects – to become brown as they age. It has been suggested that the degree of colour change may be linked to the presence and proportion of other metals, such as zinc and copper. In some drawings, inks from different batches that once matched, may now exhibit very different colours (Ambers, Higgitt and Saunders 2010: Figure 6.8).

The changes of hue seen in paint films containing smalt are the result of reactions between the pigment particles and binding medium. As we saw in Chapter 2, smalt is a blue, cobalt-containing glass commonly used by European artists in the sixteenth to nineteenth centuries, which owes its colour to absorption by tetrahedrally co-ordinated Co^{2+} ions. In common with many other glasses, smalt contains potassium, which comes from the flux used in the manufacture of the base glass. Reaction between potassium and components of the binding medium, for example, the fatty acids in oil or egg, can have two effects on the colour of the paint film. First, as potassium is lost from the glass, there is a gradual shift from tetrahedral to octahedral co-ordination of the Co^{2+} ions, resulting in loss of colour (fading). Concurrently, the reaction between potassium ions and the binding medium may generate coloured species. As a result of concurrent loss of the blue colour and yellowing of the medium, passages painted with smalt can often appear beige or brown (Figure 5.8).

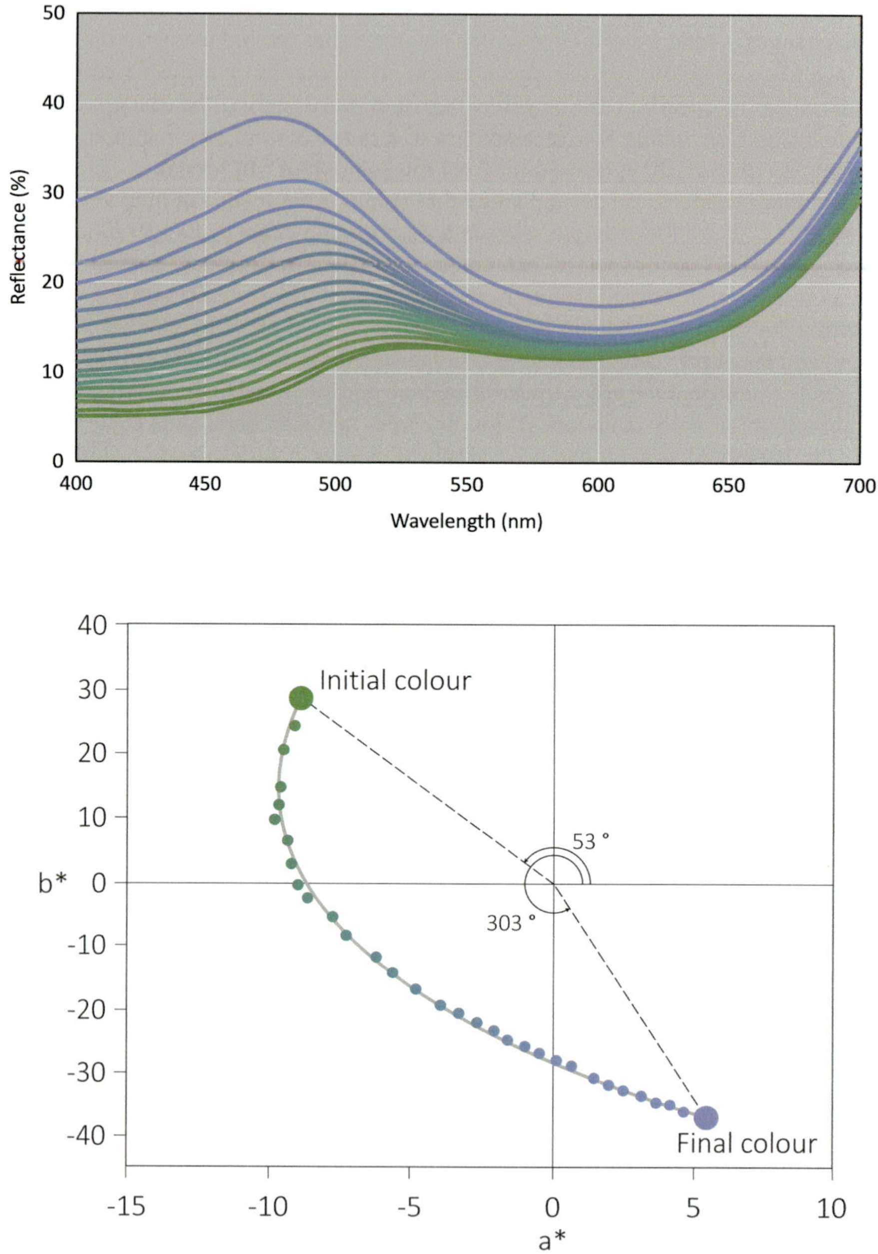

Figure 5.7 Hue shift in a mixed green pigment in which the yellow component is lost. The upper part of the figure shows reflectance spectra recorded as the mixture changes, with a shift in the maximum reflectance from the green to the blue region of the visible spectrum. The lower part of the figure shows the same changes colorimetrically, see text

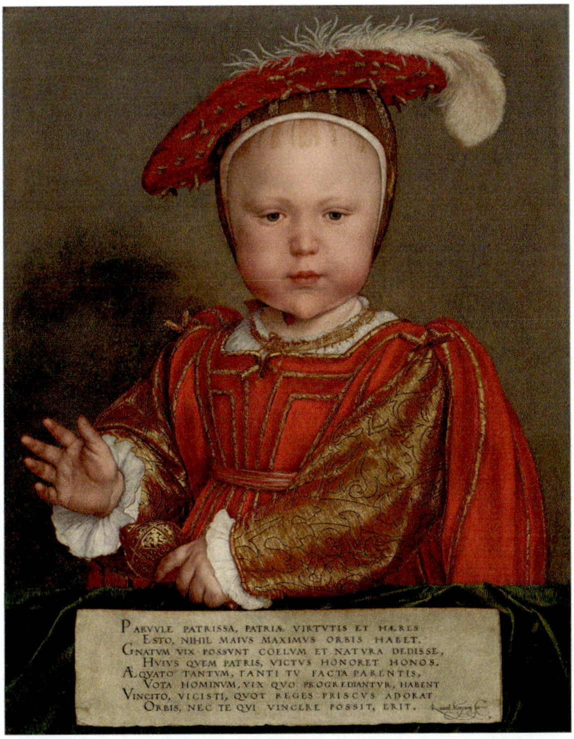

Figure 5.8 Portrait of *Edward VI as a Child* by Hans Holbein the Younger. The brown background was probably once a slate blue, as microscopic examination revealed the pigment used in this area to be smalt. [Source: Oil on panel, 1538. 56.8 × 44 cm. National Gallery of Art, Washington, 1937.1.64. Andrew W. Mellon Collection.]

A second source of shifts in hue is the alteration of one component in a mixture. Such changes occur quite frequently where fugitive red, yellow or sometimes blue organic colorants (typified by those illustrated in Figures 2.18–2.20) have been mixed with a more stable colorant.

We have already noted that natural red and yellow dyes in tapestries frequently fade, leaving the (relatively) stable blue areas intact. However, many of those blue areas may originally have been green, achieved by overdying the blue with a yellow dye (from a plant source, including weld, dyer's broom and unripe buckthorn berries), before the latter faded (see Figure 5.2). A similar situation arises in green paint mixtures based on, for example, mineral blue and yellow lake pigments made from these dyestuffs. The loss of the fugitive yellow renders once green areas blue, a phenomenon often noted in Dutch still life, flower and landscape paintings (Kirby and Saunders 1998) (Figure 5.9), and even more frequently in eighteenth-century watercolours.

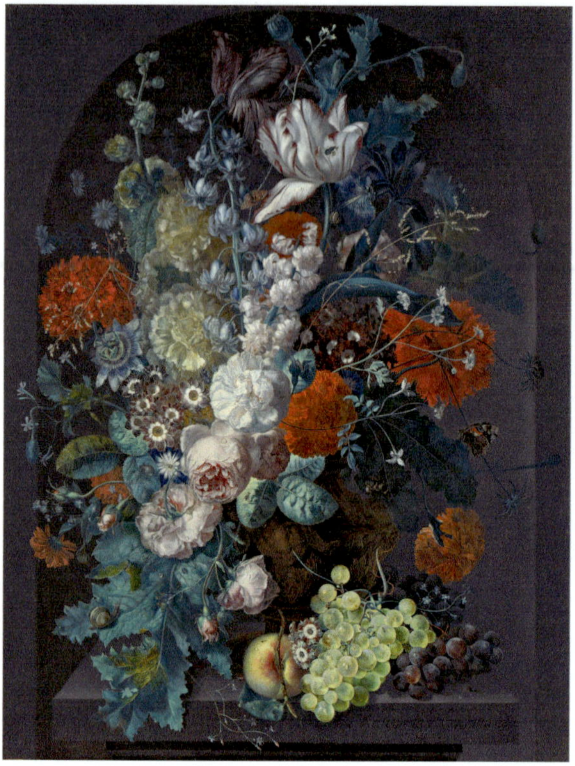

Figure 5.9 *A Vase of Flowers* by Margareta Haverman. Many of the leaves appear blue due to the loss of a fugitive organic yellow pigment from the green paint mixture. [Source: Oil on wood, 1716. 79.4 × 60.3 cm. Metropolitan Museum of Art, 71.6. Purchase, 1871.]

A similar situation can arise when fugitive red dyestuffs or lake pigments are mixed with stable blues to give a purple. As the red fades, the hue shifts from purple towards blue (Figure 5.10). Such changes in mixed purples have been reported in instances as disparate as the Virgin's robe in the late fourteenth-century *Coronation of the Virgin* by Lorenzo Monaco to the irises in Vincent van Gogh's 1888 landscape *Field with Irises near Arles* and 1890 *Irises*.

In the embroidered dress worn by Elizabeth I in the so-called Darnley portrait, areas that were probably originally purple now appear brown, due to the deterioration of both components in a mixture, with the red lake fading and smalt becoming brown as described above (Rae and Sheldon 2016).

Earlier, we saw that red lakes are among the pigments that tend to fade towards white, but it has been observed that certain materials, notably brazilwood lakes, deteriorate by way of a yellow intermediate (Peggie et al. 2018). In Figure 5.11 we can see that as redness (a*) is lost by a brazilwood lake sample, it initially becomes

Figure 5.10 Mixtures of azurite, lac lake and lead white pigments. The upper series has been kept in the dark, while the lower series has been exposed to light, causing a shift in the hue from purple towards the blue of the more stable azurite

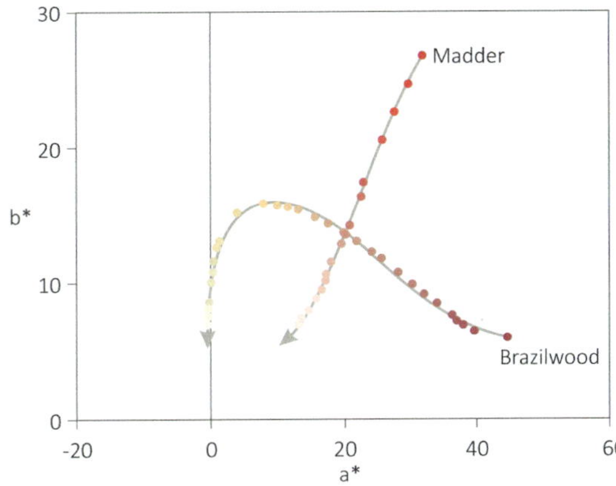

Figure 5.11 Plot of a*against b* for the fading of madder and brazilwood lake pigments [Source: Adapted from Saunders and Kirby (1994).]

more yellow before this intermediate species is also lost. In contrast, a madder sample shown alongside loses redness and yellowness (b*) concurrently.

More recent examples of hue shifts can be found in colour photographs and inkjet prints; the loss of one or more of the three coloured inks or dyes can change the colour (Figure 5.12).

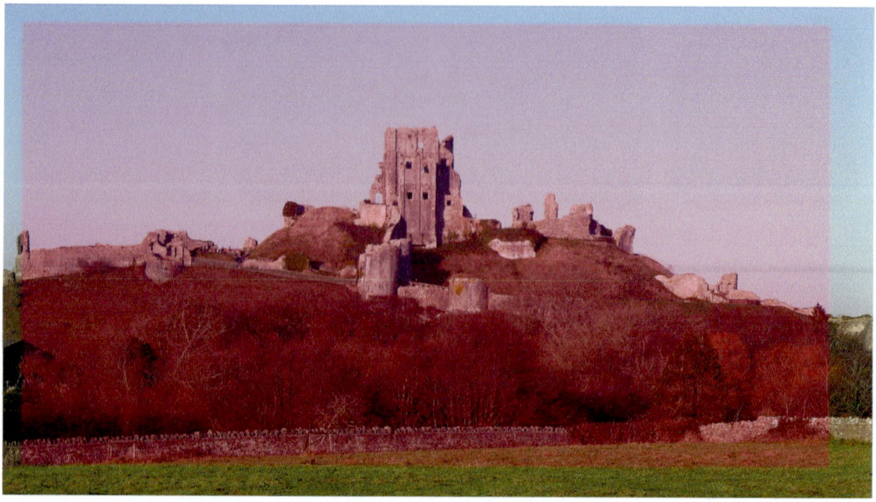

Figure 5.12 A faded digital print. The yellow and cyan dyes have faded more severely than the magenta dye, imparting a pinkish cast to the centre of the image where it has not been protected from light

5.5 Changes in transparency

We can find instances of increased or decreased transparency in materials, although examples of the latter tend to be more common. Many of the factors that affect transparency will also influence gloss, for example, dirt or dust deposited – and sometimes subsequently imbibed – at the surface of an object.

As transparency is dependent on the passage of light through an object or layer, anything that alters the balance between light transmittance and its scatter or absorption will affect transparency. Many of the processes described in Section 5.1 that cause colours to appear lighter as a result of increased scatter will also affect transparency, including abrasion, the development of a fine crack network, delamination or the loss of cohesion between particles and medium.

Chemical changes that can reduce transparency include surface corrosion or pitting, as seen in much archaeological glass (Figure 5.13), or the crystallisation of components in a once transparent layer. Whatever the origin of this increased scatter, we saw in the section on lightening that it can have the effect of desaturating – or in extreme cases masking – any colour beneath.

Conversely, a decrease in scattering can increase the transparency of a material or layer and – depending on the nature of the underlying material – may reveal or intensify the surrounding or underlying colour. A familiar example of this effect is the resaturation of a surface by the application of a coating or varnish. The difference in refractive indices between the air and the uncoated material is large

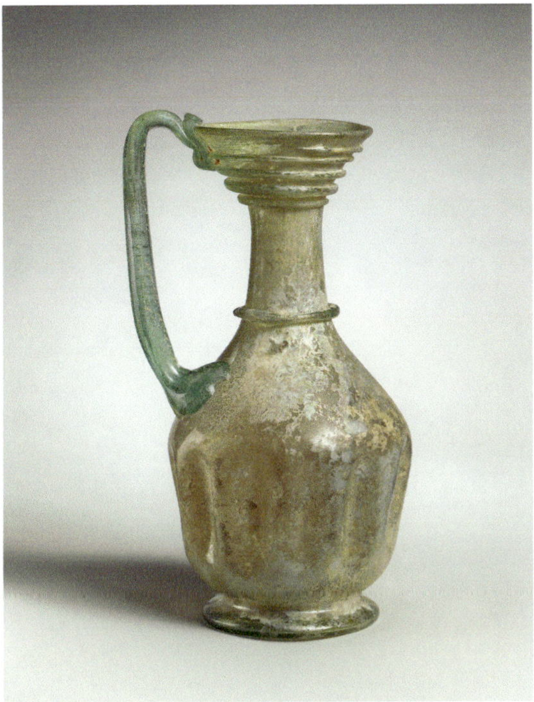

Figure 5.13 Translucent pale blue–green glass jug. Burial has resulted in surface deterioration, including pitting, iridescence and a cream-coloured encrustation. [Source: Roman, fourth–fifth century CE. Height 14.9 cm. Metropolitan Museum of Art, 41.75. Gift of the Madeline F. Astor Dick Estate, 1941.]

enough to cause considerable scatter or diffuse reflectance at a rough interface, while the refractive index difference between the underlying material and the varnish is much smaller, reducing scatter at this interface, while the varnish-to-air interface is much smoother and hence produces less diffuse reflection (see Figure 4.14).

Some chemical changes in the binder can cause increased scattering, for example, as particles separate from the surrounding medium, but other changes may increase the transparency of a material or layer. As oil paint films dry, the refractive index of the binder increases, becoming closer to that of some pigments, and scatter at the pigment–binder interface decreases (see Figure 1.8). As a result, the film becomes more transparent and may reveal the colour and form in underlying layers (Figure 5.14). We have also noted in Section 5.1 that the formation of metal soaps in paint films can have a similar effect on transparency as these compounds have refractive indices closer to that of the binding medium than the pigments from which they formed.

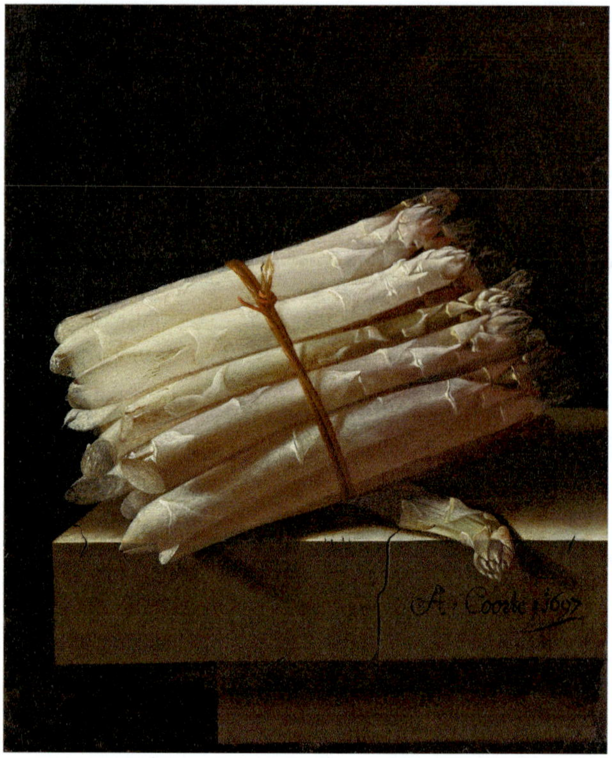

Figure 5.14 *Still Life with Asparagus* by Adriaen Coorte. The transparency of the white paint used to depict the asparagus spears has increased, revealing the underlying black background paint at their cut ends. [Source: Oil on paper attached to panel, 1697. 25 × 20.5 cm. Rijksmuseum, SK-A-2099. A.A. des Tombe Bequest, The Hague.]

5.6 Changes in gloss

As the gloss of an object is related to the ratio of specular to diffuse reflectance at its surface (see Figure 1.6), chemical or physical processes that affect the roughness and scatter of a surface will alter its gloss. We have already seen many of these changes in the context of changed lightness and transparency, including abrasion or erosion of a surface, dust accretion and chemical or physical changes that cause varnishes and coatings to degrade or 'sink' into underlying absorbent layers. To an extent, these processes may be reversible, as dust can often be removed from surfaces relatively easily and failed varnishes and coatings are often designed to be replaced periodically.

Increased gloss may arise from wear, for example, repeated touching of objects on open display, or from grease and other substances absorbed during such touching (which is often accompanied by a darkening, since the greasy surface may trap dirt). Much research has been devoted to preserving matt surfaces during conservation treatment, by avoiding over-polishing, selecting appropriate varnishes and coatings and ensuring that consolidants do not increase the gloss of, for example, pastel, chalk or charcoal drawings.

5.7 Loss of fluorescence

The appearance of those objects that owe all or part of their colour to fluorescence may be affected by chemical changes to their colorants or substrates. The deterioration of fluorescent material is not uncommon, as the absorption of radiation that triggers fluorescence or phosphorescence creates an excited state (see Figure 2.22) that could undergo a further reaction rather than emit radiation of longer wavelength. In fluorescent colorants such reactions may produce species that do not fluoresce, or that fluoresce at a different wavelength, altering the perceived colour (which is usually the result of a combination of reflectance and fluorescence). A gradual diminution in the fluorescence from optical brighteners in modern papers and some fabrics has also been noted, allowing the slightly yellow hue of the ageing substrate to become more prominent.

It is also worth bearing in mind that when displayed in museum conditions that routinely exclude ultraviolet radiation, fluorescent objects may not appear as they were originally intended.

5.8 Significance and acceptability of colour change

We have seen that it is possible to measure colour change and – to an extent – determine whether a particular alteration will be visible to the viewer under favourable conditions (the just noticeable difference, JND). In addition, the visibility of a colour change is linked not only to its magnitude but to the nature of the alteration (we are generally most aware of shifts in hue) and the context of the change, as we are particularly sensitive to changes that affect colours that we feel we 'know', such as those of familiar objects, green leaves or faces – as in other circumstances we may only be aware of a colour change as a result of having some direct point of comparison, such as an unaffected area or an unchanged analogous object, see Figure 4.3 and Figure 5.15.

Even if we can relate a colour change to a specific number of JNDs as a means of determining how visible it might appear, this does not address the question of whether such change is significant or acceptable. These criteria are much more subjective and vary among objects and collections. For instance, colour will be less crucial for the interpretation or appreciation of some objects than others and

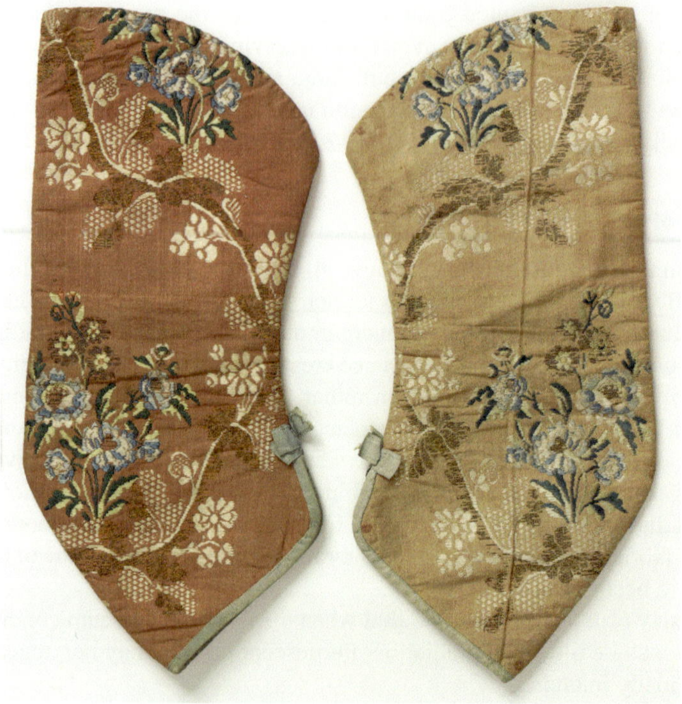

Figure 5.15 A pair of silk sleeves, for use with an eighteenth-century dress. The sleeve on the right has faded badly while the other sleeve is relatively well preserved. [Source: French, eighteenth century. *c*.38 x 17 cm. Cooper Hewitt, Smithsonian Design Museum, 1896-11-1-a, b. Gift of Mrs. James O. Green.]

while in some cases – for instance, in the patination of metals – we may see change as a positive sign of age or authenticity, even small changes in other objects, for example, pristine material, might be viewed negatively.

So, although the principle of using JNDs as a metric to assess the extent of colour change may be helpful, it is trickier to decide a threshold at which it affects utility, intent or interpretation, or can be termed damage. Indeed, Royce, Baars and Viles (2023) note that not all change is necessarily damage, that definitions of damage are context-dependent and complex, and propose a definition that attempts to reflect this complexity: "damage is a material change which is perceived to have negatively affected an item's value-defining aspects".

Whatever we feel about change or damage, there are many instances where we cannot halt entirely the physical or chemical processes responsible for them, and so must manage deterioration to prolong the lifetime of, and access to, objects.

In these cases, we generally define the time over which certain changes are considered acceptable, often expressed for convenience as a number of JNDs over a particular period. Setting such a 'lifetime' is a matter for institutional discussion and decision-making rather than scientific analysis. Evaluations of this type are familiar to those looking at colour changes caused by light exposure, where the ranking system for vulnerability described in Section 6.5 is combined with a judgement on acceptable rates of change to develop a policy for light exposure that balances accessibility and preservation (Saunders 2020: 272–277; Pesme and Dunne 2022).

5.9 The reversibility of colour change

There are a few examples of colour changes that can be reversed without chemical or physical intervention, for example, the frequently cited 'recovery' of colour by Prussian blue. In this case the loss of the blue colour by light-induced reduction can be (largely) reversed by dark storage, which allows slow oxidation partially to restore the blue colorant (Rowe 2004; Lerwill et al. 2015). Colour change in certain minerals and gemstones can also be reversed by dark storage (Nassau 1992: 20). Conversely, as mentioned in Section 5.3, the yellowing of oil paint films (and some textiles) in the dark can be reversed, or at least reduced, by exposure to strong light (Townsend et al. 2011).

However, most of the colour changes we encounter in collections are irreversible; the colorant molecule has undergone an irreversible chemical reaction, or the physical structure of the object has been altered permanently. In such cases, attempts to return to the original colour require some form of intervention. A decision to intervene is a matter not only of practicality – do we know the nature of a lost colour, and can it be recovered? – but of aesthetics and ethics – are we certain that we understand the original appearance, including modelling and shading, and will our intervention potentially cause damage, now or in the future?

Matters are a little more straightforward if we are removing surface accretions that have not become too deeply ingrained. In recent years the removal of industrial-era dirt from robust surfaces such as stone has benefitted from the introduction of laser ablation to complement more traditional methods using solvents or poultices. Dirt that has caused darkening by imbibition in a sacrificial coating or varnish can be removed along with that layer. Care is, of course, needed that the removal process does not damage underlying layers chemically or physically, such as through abrasion.

The darkening or yellowing of varnishes and coatings is a sign that they are fulfilling part of their function, to absorb dirt and radiation that might otherwise damage the underlying layers. Although this colour change is irreversible, its effect on the appearance of the object can usually be reversed as they have been designed – hopefully – to be removable without adversely affecting the object to which they were applied.

Some materials that form on the surface of objects, such as patinas, crystalline deposits or the products of pigment–binder interactions, are less easily removed. A decision is needed whether any resulting improvement in appearance is outweighed by potential alterations to the underlying layers, for example, the physical removal of a patina or its chemical treatment (often reduction) may recover the metallic colour of the object but may render the exposed surface more likely to repatinate. It also needs to be borne in mind that age patination is sometimes considered a positive attribute, see Section 5.8.

Although chemical treatment of changed materials can be very effective, it is often considered too interventive. In earlier sections we saw that lead white can darken either by oxidation or by reaction with reduced sulfur, producing brown lead(IV) oxide (PbO_2) or black lead(II) sulfide (PbS) respectively. The disfiguring nature of these changes (see Figure 4.9 and Figure 4.10) has prompted numerous attempts to treat blackened lead white (Lussier and Smith 2007). Drawings and watercolours have long been treated with hydrogen peroxide (H_2O_2), which oxidises black PbS to white lead sulfate ($PbSO_4$) (McFarland 1997). Although the final product is the sulfate ($PbSO_4$) rather than the original basic carbonate (lead white: $2PbCO_3{\cdot}Pb(OH)_2$), few deleterious side effects have been observed (although damage to the paper substrate of a watercolour thus treated would be a concern) and it has been noted that the sulfate is less likely to blacken in future.

In wall paintings, H_2O_2 is also used, but in this case acts as a reducing agent to convert the lead(IV) species in brown plattnerite (PbO_2) to lead(II) in acidic conditions. If acetic acid is used to decrease the pH, white lead(II) acetate is produced, which reacts with atmospheric carbon dioxide to form lead(II) carbonate (Matteini and Moles 1981). As there are different lead carbonates, the end product may not be identical to the original, but the process has been used for some decades and the results monitored to look for adverse side effects.

The yellowing of cellulosic materials has sometimes been counteracted by exposing paper or textiles to strong light (e.g., in the traditional light bleaching of linen) or by controlled aqueous washing to remove the discoloured material. In recent years the latter process has been adapted using gels and other methods that allow greater specificity and control. Care is needed that other substances, including colorants, are not mobilised by washing or affected by bleaching. As some of the discoloured materials are acidic, washing has the added benefit of removing compounds that could promote further deterioration.

Finally, a relatively recent concept is the recreation of original colours through coloured lighting. A method for reversing the yellowing effects of an aged varnish using 'blue' lighting was proposed by Lafontaine (1986), but the advent of data projectors has allowed patterns of coloured light to be projected onto objects that 'restore' the supposed original colours without physical or chemical intervention.

Such techniques have been applied in temporary displays to reintroduce lost colours in tapestries (Perkins et al. 2011), and to compensate for the fading of lithol red in paintings by Rothko (Stenger et al. 2016).

Finally, although living organisms frequently regenerate lost colorants, these processes cease after death, so that colour changes in natural history specimens and materials derived from plants and animals are generally irreversible.

6 Why we measure colour

This chapter covers some of the most frequent reasons for which colour is measured in the cultural heritage sector, as well as some of the requirements, issues and uncertainties associated with each of these types of study of colour or colour difference.

On some occasions we may wish to use colour measurements as a way of identifying the materials found in objects (Section 6.1). This, as we will see, is not always easily achievable, but even when an unambiguous identification is not possible using colour measurement alone, mapping the colour across an object may provide a useful indication of the distribution of a colorant (Section 6.2), which can be combined with other analytical methods appropriate to its characterisation.

More frequently, colour is measured to make comparisons between objects or samples, often after the passage of time or after an object has undergone conservation treatment. Sometimes these measurements are made directly on objects to quantify or monitor changes in colour (Section 6.3), while other studies use model or surrogate samples as a means of predicting the changes that may occur to objects (Section 6.4).

6.1 Identification of materials

The visible spectra of many coloured materials contain very broad bands, for reasons discussed in Sections 2.1.1.3 and 2.1.1.4, which makes material identification challenging. Some classes of colorants have characteristic peaks in their reflectance or absorbance spectra, for example, cobalt-containing glasses, glazes or pigments and dyestuffs or lake pigments containing various anthraquinones. Identifying colorants requires good spectral resolution, as peaks can be 'lost' if the wavelength interval between measurements is too great (Figure 6.1). It is sometimes possible to differentiate between potential colorants by the position of their peaks or absorption edges (an abrupt increase or decrease in absorption as wavelength increases), but this also requires good spectral resolution. As colorants with similar molecular structures can give rise to peaks at similar wavelengths, accurate wavelength calibration is important in analytical studies.

Even with sufficient resolution, it can be difficult to establish the maxima for minor peaks. Calculating a so-called first derivative spectrum is a useful method

DOI: 10.4324/9781003397182-7

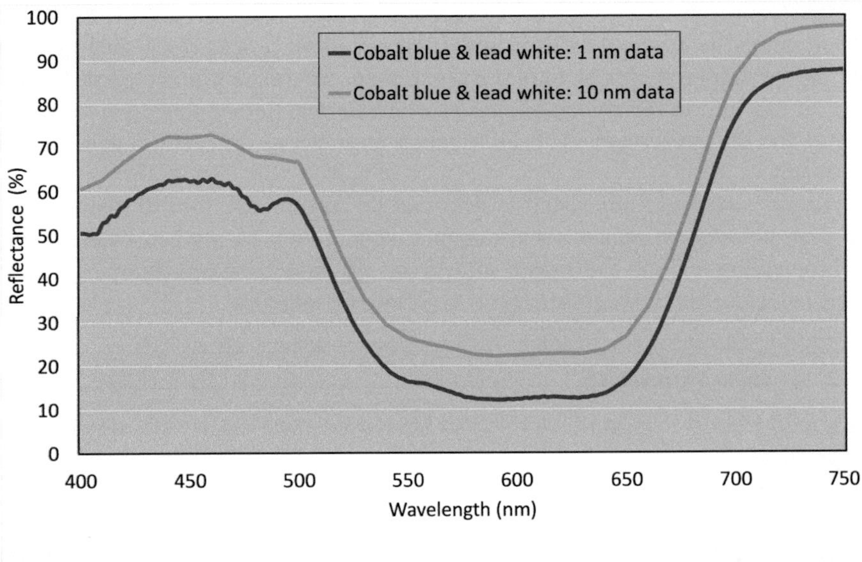

Figure 6.1 Reflectance spectra of a mixture of cobalt blue and lead white recorded at either 1 or 10 nm intervals. In the latter spectrum the peaks are much less well defined. The two spectra are offset for clarity

of determining the position of maxima or absorption edges. A good example is provided by a study of the anthraquinone-based pigments madder lake and cochineal lake, which used FORS with a spectral resolution of 1.4 nm and first derivative spectra to produce a decision tree to help identify these common red organic pigments – although care is needed, as the bands can be shifted according to the nature of the substrate (Fonseca et al. 2019; esp. Figure 2).

For most colorants, however, the visible spectrum provides insufficient information on which to base unambiguous material identification and it is common to extend the range of wavelengths into the infrared region, which often contains further absorptions characteristic of the molecular structure of the compounds. Libraries of reference spectra for materials found in works of art (many acquired using FORS) can then be used to identify the colorants present (Cosentino 2014; IFAC 2021).

In addition, materials that fluoresce can sometimes be identified by an analysis of their excitation–emission spectra. Such a technique has proved useful in the identification of traditional occidental red lake pigments (Nabais et al. 2018).

Colorimetric measurement is not a good tool for colorant identification. While colorimetric data unambiguously identify a colour, they do not give information on the colorants from which it is produced as the same colour co-ordinates can arise from different pigment mixtures.

6.2 Mapping the presence of materials across an object

Determining the distribution of materials across an object can often shed light on the use of colorants, for example, if more than one rather similar pigment was used, it can inform the choice of sites at which to apply other analytical methods, or provide the conservator with information on material distribution that can aid treatment.

As we have seen, our visual system can be an effective tool to discriminate between colours, so image-based mapping should only be applied where it can add something that the eye cannot see. As we saw in Section 3.9, mapping can be achieved using either spectrometric or colorimetric imaging.

6.2.1 Spectrometric mapping

Imaging spectrometers, particularly those developed in the last decade, are powerful reflectance spectrometers capable of making millions of individual measurements of colour across an object's surface. When used in the visible region, imaging spectrometry (hyperspectral imaging) can provide a nuanced mapping of different colours, as a (frequently high resolution) spectrum is available for each point on the surface.

The mapping can be based on single wavelengths characteristic of specific colorants or on the transformation of the spectral information into more manageable data. In the visible region, it is common to convert the spectral data to CIE colorimetric data, as described in Chapter 3, but other computational methods, including principal component analysis (PCA), are often applied, particularly when the data span the visible and infrared regions. The latter technique was successfully applied, for example, to differentiate between iron gall and carbon-based ink in the *Gough Map of Great Britain* (Huang, Messinger and Bove 2020).

Visible imaging spectrometry is less useful if we wish to identify the colorants and other materials present within the areas delineated in such maps. While electronic transitions in the visible region give rise to colour, it is the vibrational transitions in the infrared region that tend to be characteristic of molecular structure, revealing invisible differences in the materials employed. An example is offered by the examination of *Lavender Mist* by Jackson Pollock, for which it was possible to map the distribution of oil- and alkyd-bound white paints (Dooley et al. 2017, Figure 3d).

Applications of imaging spectrometry are becoming increasingly sophisticated. For example, recent studies have used fluorescence imaging spectrometry to map the distribution of eosin in Van Gogh's *The Olive Orchard* (Dooley et al. 2020), and a combination of reflectance and luminescence imaging spectrometry to examine a Roman Egyptian funerary portrait (Radpour et al. 2022). Visible and infrared imaging spectrometry – and its analogue in the X-ray region of the electromagnetic spectrum – have until recently been confined to the research laboratory but are increasingly available to researchers and practitioners in the cultural heritage field as equipment becomes less expensive and more user-friendly.

It is worth noting that using wavelengths outside the visible introduces the issue of transparency. As wavelength increases, scatter decreases, so images made in the infrared often contain information about the reflectivity of layers deeper in the structure of an object. Although this can be potentially confusing or misleading, it can also be advantageous, for example, in paintings and drawings, it may allow earlier underlying versions or preliminary drawings to be seen.

6.2.2 Colorimetric mapping

The advantage of colorimetric mapping is that it is generally accomplished using systems based on commercially available – and hence much less expensive – colour cameras adapted, as described in Section 3.9, with filters or specialised lighting, and is more easily implemented than spectral imaging methods; a useful summary is given by Kuzio and Farnand (2022).

As we saw in Chapter 3, imaging colorimeters do not generally produce data that can be used to identify materials or to detect small changes in colour. The images have sufficient accuracy, however, to allow segmentation of the image into regions with a similar surface colour, which in some cases is sufficient to identify patterns of material use (Figure 6.2). Such maps can also guide more detailed investigations, highlighting areas that may contain the same colorants, so that results from analytical methods applied at one location can be (cautiously) extrapolated to other areas.

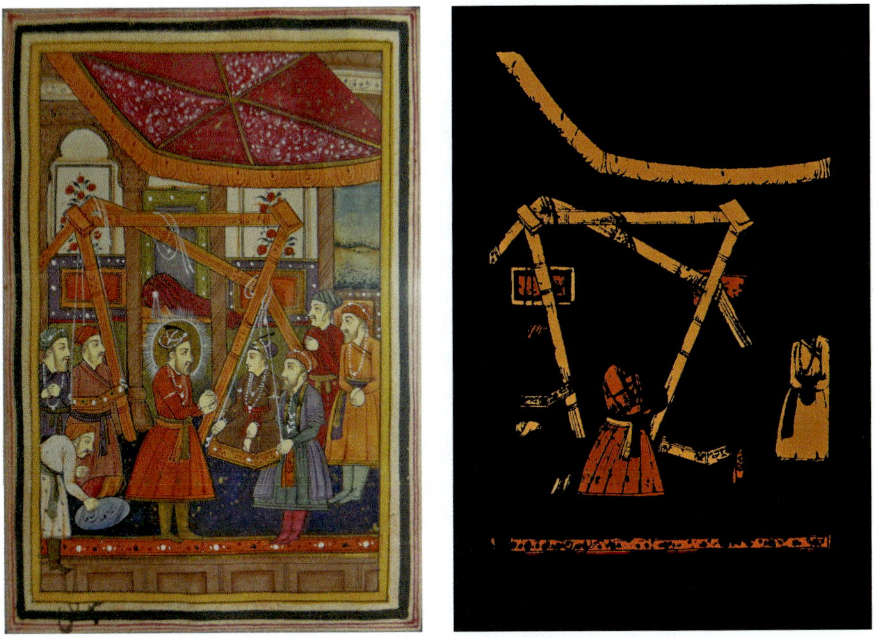

Figure 6.2 An Indian miniature (left) and a map (right) showing the distribution of two colorants according to their colorimetric properties

Moreover, the colour information in colorimetric images is often used in high quality printing or manipulated to produce visualisations of former or future states of an object. Such visualisations have often been used with objects altered by light-induced colour changes. Examples include the simulation of the appearance of Pedro Campaña's *The Conversion of the Magdalen* before its red lake pigments faded and passages of smalt browned (Richardson, Saunders and Spring 2001) and the visualisation of the original appearance of the foreground vegetation in Georges Seurat's *La Grande Jatte* before the lead chromate pigments it contains darkened (Berns et al. 2005).

Berns and his colleagues have also used colorimetric images to visualise the future appearance of objects, so that the effects of different degrees of light-induced deterioration can be appreciated – and hopefully minimised through appropriate preventive conservation policies. For Vincent van Gogh's *Bedroom at Arles*, colorimetric imaging was used – in conjunction with spectral information collected on certain areas – not only to simulate the original appearance of the painting but to predict its future appearance in response to particular light exposures (Berns 2019; Hendriks, Brokerhof and van den Meiracker 2017).

It is worth noting that although colorimetric and spectral information gives us the ability to make such visualisations, they remain subjective, speculative reconstructions.

6.3 Measuring and monitoring colour change in objects

For reasons we explored in Chapter 4, many cultural heritage materials are prone to colour change. Monitoring the colours of objects as they age, or change as the result of display conditions, helps to inform future care and can also give an impression of how an object might have appeared when created, or might appear if subjected to particular future conditions, perhaps through the visualisations described in Section 6.2.

In contrast to studies of surrogate samples detailed in Section 6.4, there have been fewer programmes to monitor colour change in objects themselves. Such studies are more difficult to conduct as care is needed to avoid damaging the surface during measurement and because the areas under examination are often less homogeneous than experimental samples, making it important to have a means of returning to the same area for repeat measurements, perhaps using a template (see below).

Monitoring colour change also relies on the availability of baseline measurements against which future readings can be compared. There are several examples of relatively short-term monitoring programmes (although their timeframe is generally much longer than sample-based experiments that tend to use accelerated ageing to produce more rapid colour changes), which often aim to determine the change caused by a particular exhibition or thematic display. For example, at the National Museum of Australia, Ford (1992) studied changes to dyed textiles during a three-month exhibition and Ford and Smith (2017) assessed the changes to colorants in various types of object after display for three years.

Longer-term evaluations of change are less frequent, principally because equipment ages in a manner that cannot be accounted for through calibration, or becomes obsolete and is superseded. As an example, the colorimetric system at the National Gallery, London (described in Section 3.5) was replaced first by a reflectance spectrometer (Bullock 1978) and then four generations of imaging colorimeters/spectrometers (Liang, Saunders and Cupitt 2005); although comparing data from successive generations of equipment proved possible, it was far from straightforward.

In both short- and long-term programmes the selection of measurement points is key and is generally informed by sample-based studies of colour change that help us understand which materials are most prone to change and, therefore, which objects or parts of objects we might monitor most closely. Site selection can also be informed by mapping areas of similar colour across the object, as described above.

Having selected measurement sites, it is essential to be able to return to that exact point after some months or years. This has been done mechanically, using a 'rig' that holds the object and allows the measuring head of the colorimeter or spectrometer to be moved to a precise recorded position. More commonly, a template that has markings to allow the exact positioning of the instrument is laid over the object. Templates are often made of an inert transparent material such as Mylar®, with holes at the measurement site to give unimpaired access to the object's surface. Increasingly, images are made to document the position of a measuring head with respect to the object (e.g., Ford and Smith 2017). Many modern point colorimeters and spectrometers have inbuilt digital cameras, while imaging colorimeters and spectrometers provide an intrinsic record of the position of measurements.

In addition to monitoring changes during display, which are largely brought about by exposure to light or ultraviolet radiation, colour measurements on objects are also used to assess the success of conservation treatments. While colour change is desirable in some treatments, for example, cleaning, on other occasions, studies on surrogate materials will likely have been made to identify possible unwelcome colour changes; in these cases, colour measurements on objects help to confirm the sample-based results.

The effect of applying consolidants or coatings to objects is sometimes monitored in this way, although most studies look only for immediate differences in colour and not at any longer-term changes that might result from the treatment, which can be difficult to isolate from the effects of display.

6.4 Investigating and predicting colour changes

While measurements on objects give a direct indication of any changes they are undergoing during routine display and storage, there is understandable reluctance to expose objects to conditions that might cause change for purely experimental purposes. To help us to predict and ameliorate colour change, we often rely on experiments that expose representative – or surrogate – samples to the conditions that might cause objects to change or to the conservation treatments we are considering applying.

Many of these experimental studies of colour change in the heritage field thus fall into one of two categories: (1) assessments of the way in which conservation treatments affect colour; and (2) predictions of the colour change caused by display conditions (principally those changes caused by exposure to light or ultraviolet radiation).

How closely experiments on surrogate materials predict the behaviour of objects depends on the choice of materials and the experimental conditions. For studies of colour change during conservation treatments, a major hurdle is sourcing test samples that mimic the behaviour of aged materials, both in the object and in any old coatings, adhesives, etc., that we might wish to remove. When the effect of applying new coatings or consolidants is studied, the situation is a little easier as we need only find an appropriate surrogate for the aged object surface, bearing in mind we must consider not only its colour, but also properties such as porosity, which will affect the colour or gloss after treatment.

The preparation of appropriate samples is usually informed by analytical studies of the objects they are to emulate, supported by an understanding of how original materials were sourced and prepared – often based on studies of historical documentary sources. Although most experiments use historical reconstructions, we are sometimes fortunate enough to have access to surrogates that are almost identical to the original; an example is offered by the conservation of Rothko's *Black on Maroon*, where it was possible to make tests using materials remaining from Rothko's studio (Ormsby et al. 2014).

The situation is rather more complex in ageing studies; here we need to consider not only how well our surrogates represent the original materials, but also the extent to which the conditions we apply simulate the ageing process. We often seek, using accelerated ageing regimes, to simulate over the course of a few weeks changes that normally might take 50 or 100 years. Much accelerated ageing explores the effect of light on colour change and relies on the **principle of reciprocity** (see *Reciprocity* box) to equate exposure to high light levels for a short period to longer periods on display at a lower level.

Reciprocity

The principle of reciprocity, known as the Bunsen-Roscoe law, states that the amount of damage caused by a particular dose of radiation is not affected by the time over which that dose is delivered.

Dose = level of exposure x length of exposure

The same damage will thus result from exposure to a light level of 20,000 lux for 100 hours (in an accelerated ageing experiment) as from an exposure to 200 lux for 10,000 hours (several years of typical display); in each case, the dose is 2,000,000 lux hours.

However, as we saw in Chapter 4, other factors can contribute to colour change, such as elevated temperature, moisture and the presence of pollutants. Increasing the rate at which light is delivered may alter the balance between light-induced reactions and these other processes, affecting the comparison between accelerated and 'natural' ageing. Analytical methods are sometimes applied to assess the similarity between the products of these two types of ageing.

Frequently, we will find ourselves combining the two categories of colour change experiment described above; first, comparing the short-term effect of different treatment materials or methodologies on colour and then assessing – through accelerated ageing – the long-term colour stability of our favoured conservation materials.

Although surrogates are not always representative of the exact substances in the object and sometimes fail to account for pre-existing ageing of materials in objects, which alters their behaviour, their study has nevertheless generated a huge body of information about how colours change, reported within the conservation literature and summarised in Chapters 4 and 5.

6.5 Ranking colour change

Once we have a method of quantifying colour change – most frequently using units of ΔE in the CIE Lab system – it can be used, *inter alia*, to rank the relative stability of colorants or as a measure of the damage caused to materials by conditions of display or storage.

6.5.1 Colorant stability

The broader colorant industry is understandably keen to market products whose colour is stable in the situations in which they will be used. For example, automobile or exterior house paints undergo standardised testing to ensure they can withstand exposure to strong sunlight and extreme weather.

Most museum objects are exposed to less severe indoor conditions and the main concern is usually radiation-induced damage from exposure to light and ultraviolet radiation. Individual colorant manufacturers often apply their own permanence rating systems, but there are also standardised, internationally recognised testing methods, a good example of which is the ASTM Standard Test Methods for Lightfastness of Colorants Used in Artists' Materials: D4303-10 (ASTM 2010). This specifies four light exposure regimes and ranks materials into five categories (I–V) based on the magnitude of colour change measured, using the CIE ΔE_{76} system. The three most stable categories – I (excellent), II (very good or good), and III (fair) – are of most relevance to practising artists wishing to ensure the longevity of their works or for conservators who need to select stable colorants to use in conservation treatments.

6.5.2 Object vulnerability

Manufacturers' permanence rankings are less useful when examining historical colorants incorporated into objects, whose precise nature and behaviour may not be known. As described above, where we can identify colorants, we sometimes use surrogate materials to examine their behaviour when exposed to light or other agents of deterioration. Because such experiments are not standardised to the same extent as industry-testing methods, we tend to use internal standards that allow us to compare the amount of colour change undergone by the surrogate sample with that of the standards.

When assessing the effect of light exposure, the most widely used such internal standards are the so-called ISO (International Organization for Standardization) **Blue Wool Standards** (BWSs) (ISO 2014a; 2014b). This scale comprises eight woollen cloths containing dyes of increasing fastness from the most fugitive (BWS1) to the most stable (BWS8) (Figure 6.3).

The degree of fading of each blue wool is sometimes assessed by relating its change in lightness to a greyscale chart, but more often a colour difference measurement is made using the CIE ΔE system. By plotting the change of colour of several of the Blue Wool Standards (BWSs) alongside that of the material we wish to test, we can categorise its fugacity in terms of the nearest blue wool standard. Figure 6.4 shows the colour change in BWSs 1–4 and two pigment samples exposed under the same experimental conditions. We can say that one sample has a similar stability to BWS2 while the other sample is more fugitive than BWS1.

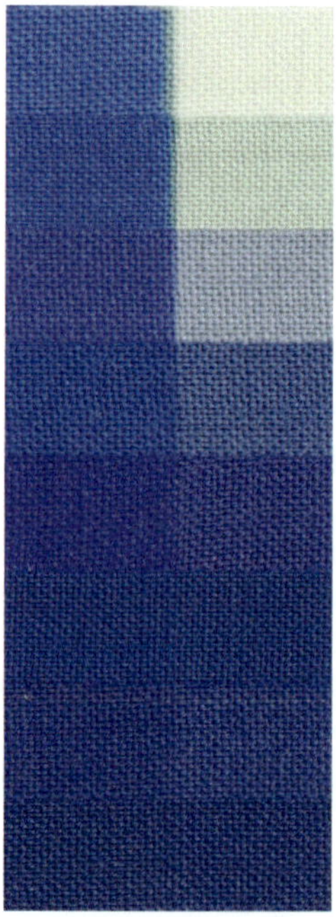

Figure 6.3 Blue Wool Standards 1–8. The left portion of each sample has been protected from light. BWS1 (top) is the most affected and BWS8 (bottom) least affected

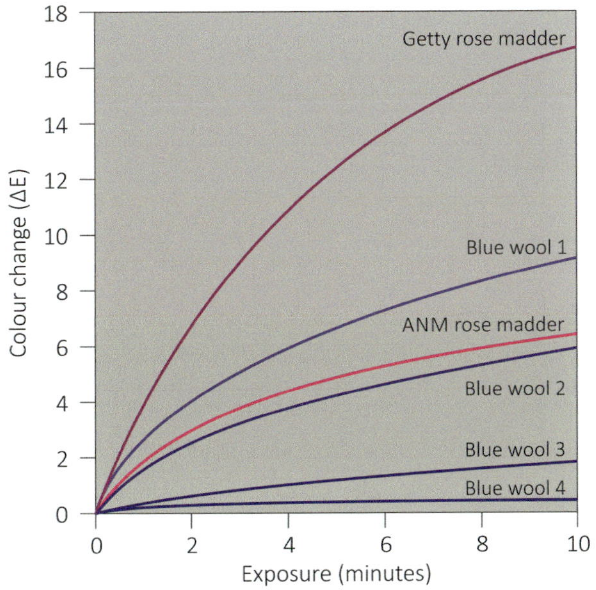

Figure 6.4 Plots of colour change against light exposure for two samples of rose madder tube paint. The graph shows the curves for BWSs 1–4 exposed under the same experimental conditions. [Source: After Ford and Smith (2010).]

Blue Wool Standards have been used in the cultural heritage sector since the 1950s, including in a landmark study of colour change in natural dyestuffs by Padfield and Landi (1966). They also form the basis of categorising the vulnerability of different materials to light exposure. Feller (Feller 1975: Table I) and Thomson (Thomson 1986: Table 22) both proposed this scale as a useful means of grading lightfastness, each suggesting three categories of vulnerability based on equivalence to a range of BWSs. Such equivalences continue to accompany current categorisations of sensitivity to light, for example, that available online from the Canadian Conservation Institute (Michalski 2018: Table 3) or in *Museum Lighting* (Saunders 2020: Table 4.5).

Some caution is needed when using ΔE to compare the colour change in samples with that of BWSs. While the change in the blue wool cloths is a relatively straightforward fading process, we saw earlier that considering ΔE alone sheds little light on the nature of the change that is occurring in some of the materials we may encounter, for example, those containing mixtures of colorants. It is also worth noting that critical properties, including physical strength, may alter without this deterioration manifesting as a colour change.

In addition, fading is not a visually linear process. The relationship between colour change and exposure to light is not linear, since successive just noticeable changes in the colour of an object will not be brought about by equal exposures to

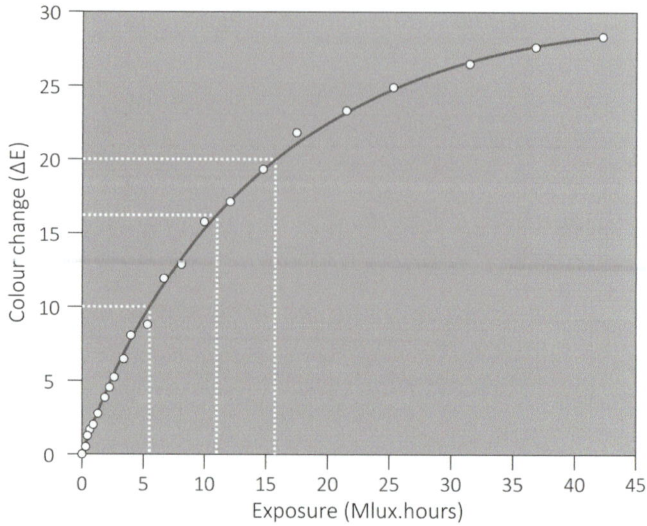

Figure 6.5 Plot of colour change against light exposure for gamboge in watercolour

light (Figure 6.5). Although they provide a numerical description of visual changes, neither measures of JND nor CIE methods of describing colour give an indication of the absolute concentration of colorants (Johnston-Feller 2001: 93).

6.6 Microfade testing (MFT)

As we have seen, attempts to quantify the effect of light by studying surrogate samples are hampered because the behaviour of these materials does not always quite mirror that of the objects they are meant to represent. It can be difficult to replicate the components and method of preparation of the original, or to account for the changed chemistry or partial fading of a material that has already aged for some years. Equally, we have been reluctant to assess the effects of light-induced or other ageing on the objects themselves.

Microfade testing (MFT) was developed by Whitmore and colleagues in the late 1990s precisely to address these issues, by allowing an extremely small area of an object, preferably in an inconspicuous region, to be used to test lightfastness (Whitmore, Pan and Bailie 1999). Although fading is carried out on peripheral areas where this is possible, some – albeit small – change to the object will occur, but this is often considered acceptable in order to gain information that helps protect the object as a whole.

The configuration of a microfade tester is very similar to that of the reflectance spectrometer shown in Figure 3.2a. The principal difference is that the incident white light comes from an extremely powerful source and is focused onto a very

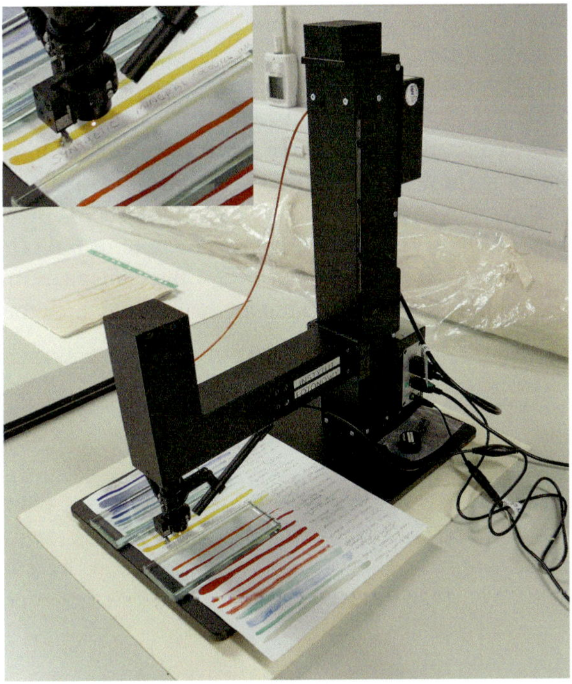

Figure 6.6 Microfade tester. The inset shows the intense illumination of a very small area of the yellow test strip, the spectrometer head and the camera used to monitor the position of fading and measurement

small region, typically <0.5 mm in diameter. This intense exposure allows a light dose that might be experienced during a decade to be delivered in under an hour.

Any colour change is measured by passing the light reflected from the sample through a spectrometer (Figure 6.6). The spectral information gathered during light exposure is typically converted by the microfade tester to colorimetric data and presented as colour change in ΔE units plotted against time. As experimental parameters vary with instrument and lamp output tends to decline over time, it is usual to expose BWSs under the same conditions and express the stability of a sample with reference to the BWS with the closest behaviour, as has been done in Figure 6.4.

While MFT equipment and practice have developed over the 25 years since its inception – for a recent summary see (Beltran et al. 2021) – the technique is best suited to assessing comparative rather than absolute light-stability. It has proved a useful tool in screening objects proposed for display, helping to inform display and exhibition policies for objects that are most sensitive to light, see Ford and Smith (2017).

6.7 Measuring the colour of light

In addition to measurements on or related to objects, conservators and conservation scientists are often called upon to measure the colour of the white light sources employed in museums or the filters with which they are sometimes used. For white light sources, we are usually interested in three factors: (1) the colour temperature of the light source (how 'warm' or 'cool' it appears, as defined in Chapter 1); (2) the colour quality of light (how well it reveals the range of colours in an object, see *Colour quality* box); and (3) the damaging potential of light (principally how much ultraviolet radiation it contains). To control the damage of materials on display, we will also wish to consider the amount of light to which they are exposed, see Section 7.2.

Colour quality

When displaying objects, we usually wish to select a white light source that can portray or *render* colours well. Methods of defining how well colours are rendered assume that the CIE standard illuminants described in Chapter 3 render colour perfectly and assess the performance of other sources in comparison to them. There are two current methods:

- The CIE **Colour Rendering Index (CRI)** is the older method, but is still widely applied (CIE 1995b). The CRI is determined by how well a white light source renders eight (or sometimes 14) colours when compared to a standard illuminant. A CRI (sometimes termed Ra as it is the average for the eight – or 14 – colours) of 100 implies a perfect rendition. Lower values indicate decreasingly poor rendering.
- The more recent **TM-30-15** measure of colour rendering, termed the fidelity index (Rf), was developed by the Illuminating Engineering Society of North America (IESNA) (IESNA 2015). It uses the results from 99 colours to compare test white light sources with standard illuminants and applies an improved model of how colours appear to the human visual system. An Rf of 100 corresponds to a perfect rendition.

In 2017, the CIE published a standard for calculating a fidelity index (Rf) similar to that proposed by IESNA (CIE 2017). In 2018, IESNA modified their standard (termed **TM-30-18**) so that the colour fidelity indices in both systems are now identical (IESNA 2018).

The colour temperature, colour quality and ultraviolet content of a white light source can be calculated from its spectral power distribution and, as described in

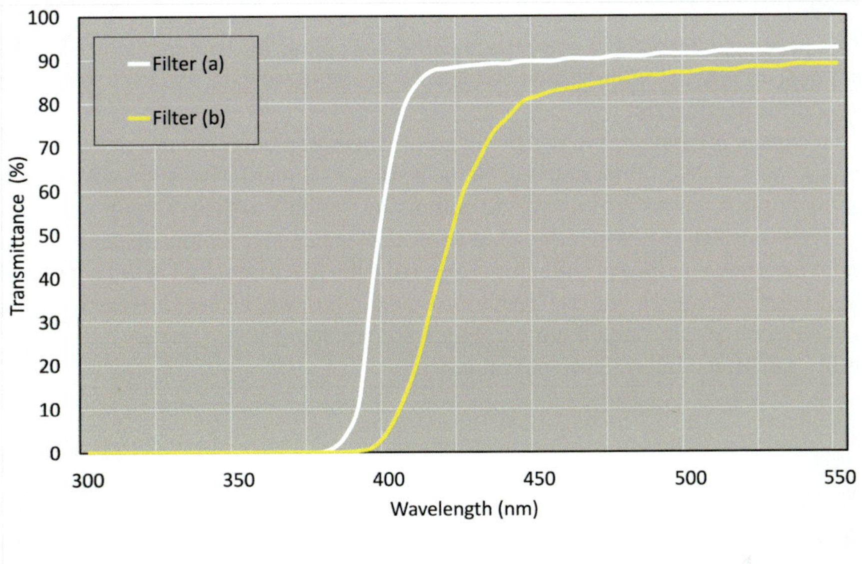

Figure 6.7 Transmittance spectra for two ultraviolet filters. Filter (a) absorbs efficiently in the ultraviolet region but not in the visible region. Filter (b) also absorbs at the blue end of the visible spectrum and consequently appears yellow

Chapter 1, spectroradiometry can be used to measure that spectral power distribution directly. If we do not have access to spectroradiometry, published spectral data can be used to calculate these parameters. However, as these calculations are quite laborious, we might prefer either to trust the data provided by lamp manufacturers or use an online resource, such as that offered by the National Gallery, London (Padfield n.d.), which contains information on many light sources and can calculate colour parameters from spectral information provided by the user.

When we employ filters that are designed to block ultraviolet radiation (in order to reduce damage to objects on display), we generally wish to reassure ourselves that while ultraviolet is indeed eliminated, not too much radiation at the blue end of the visible spectrum is absorbed, as this can increase the yellowness of the transmitted light. As an example, two transmittance curves are presented in Figure 6.7. Filter (a) absorbs in the ultraviolet region of the spectrum up *c*.400 nm, but transmits visible wavelengths above this threshold and will not alter the colour of the light transmitted. While filter (b) absorbs ultraviolet effectively, it also absorbs light in the blue region of the visible spectrum so that the transmitted light will lack the blue component and appear yellow (the colour complementary to the radiation that has been removed).

6.8 Accuracy, repeatability and reproducibility

How **accurate**, repeatable and reproducible our colour measurements need to be will depend greatly on the use for which we intend these data. We can first define these three parameters:

Accuracy is the degree to which our colour measurements conform to the correct values. We can determine accuracy by measuring a standard whose spectral, colorimetric or gloss values are known and comparing our data to these known values.

Repeatability expresses the likelihood that successive colour measurements under the same conditions will return identical data. It is often used in the context of short-term differences in measurements, for example, when several measurements are taken on the same area. Figure 6.8 shows examples of how accuracy and repeatability can affect colorimetric measurements.

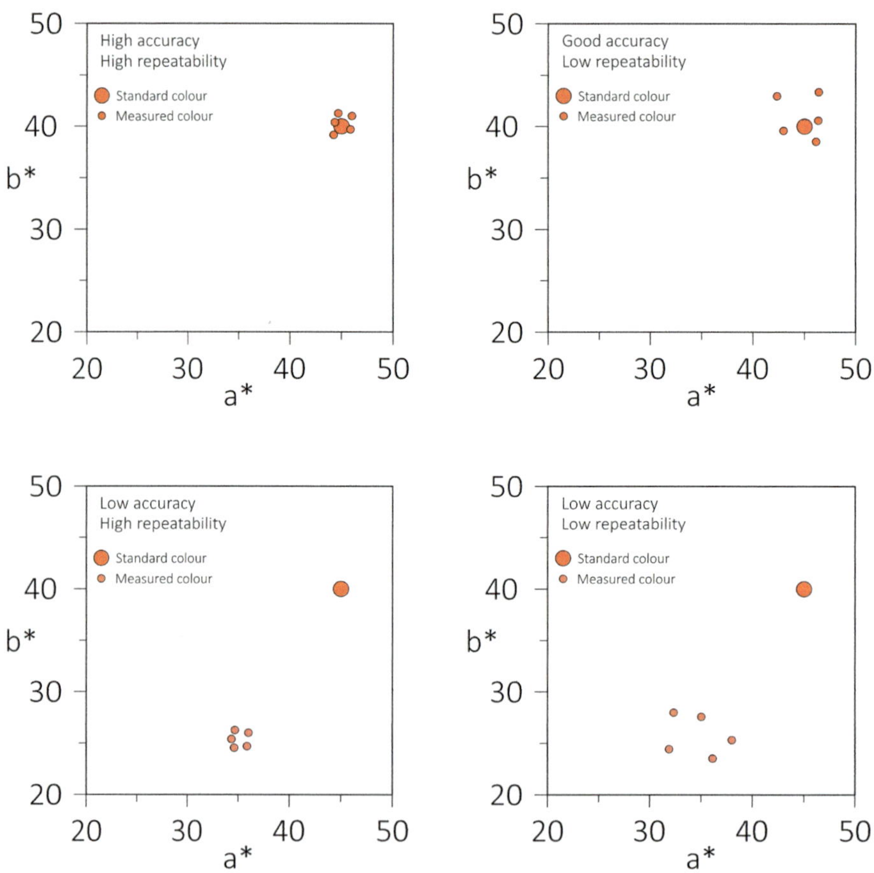

Figure 6.8 Examples of the effect of accuracy and repeatability on the difference between a set of measurements and the standard colour

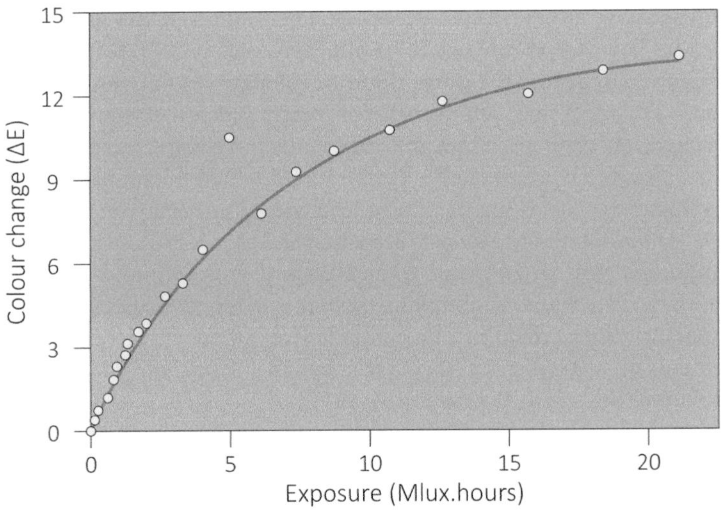

Figure 6.9 Plot of colour change against light exposure. In the context of the many data points recorded, the rogue reading at 5 Mlux hours is easily identified

Reproducibility expresses the likelihood that a colour measurement performed using the same methodology will produce the same colour data as an initial measurement. It is similar to repeatability, but more frequently used in instances where measurements are made at different times or in different places.

Colour measurements can be affected by both systematic errors or random errors in the instrument or measurement procedure. Systematic errors include changes in the instrument (often termed 'drift'), which can be ameliorated by frequent recalibration, or different methods of working by different users. Systematic errors can affect either the accuracy of the measurement or – in the case of changed working practices – their reproducibility, which can be important in long-term experiments.

Random errors may arise from, for example, electronic noise in an instrument, a change in environmental conditions (particularly temperature) at the time of measurement or incorrect calibration of the instrument for a particular set of readings. Increasing the number of measurements can allow some random errors to be spotted, either by making more readings on a sample or taking more frequent readings during an experiment. For example, in Figure 6.9 the reading taken after 5 Mlux hours does not follow the pattern of other measurements and can be discounted as a random error.

6.8.1 Accuracy

We have seen that it can be challenging to identify materials from their visible spectra. If we are to have any success, we need to be certain that the values for any peaks or absorption edges are accurate. Two important factors will be good wavelength calibration and sufficient spectral resolution. If the wavelength scale of a spectrum is incorrect, the positions of peaks or edges will be misrepresented. As the examples in Figure 6.1 show, if the wavelength resolution is poor, the peak maximum wavelength may be misread. For material identification, the accuracy of the reflectance, absorbance or transmittance tends to be less crucial, but if we wish to make an absolute measurement – perhaps with which to compare a subsequent reading – then we must ensure that the reflectance, absorbance or transmittance are calibrated against a known standard, see Chapter 3.

An important step in ensuring good accuracy in colorimetric measurements is appropriate instrument calibration, typically using a white (and sometimes black) standard as described in Chapter 3. This is often supplemented by making measurements on a broader set of colour standards, for example, the BCRA–NPL Ceramic Colour Standards (Malkin et al. 1997).

When we use colorimetry for mapping, accuracy can be important to differentiate between colours arising from the use of different colorants. As colorimetry is not generally helpful for material identification, this technique is used predominantly to look at how colours change. Although this implies that we should pay attention to repeatability and reproducibility, accuracy continues to be important, particularly if the colour differences that we are expecting to encounter are small.

6.8.2 Repeatability

Taking our definition of repeatability as a short-term source of error, we will be concerned principally with: variations in the performance of the spectrometer or colorimeter; the homogeneity of the colour under investigation; and inconsistencies in the procedure that we apply to measure this colour. As we tend to test these parameters by making a sequence of readings, we can assume that the instrumental settings have not been altered between measurements.

To test instrumental repeatability, it is common practice to make a series of measurements on the same point on an object or standard without moving the equipment between readings. This will give an indication of the effect of, for example, instrumental noise or fluctuations in the room environment on the data.

Assessing the colour homogeneity can give an indication of the extent to which it will be necessary to return to the exact position to make a further measurement, i.e., the importance of repositioning. If several readings are taken, each time moving the measuring head slightly within an area of colour, their differences can be compared with those made when investigating instrument noise. As well as highlighting inhomogeneities, these readings will indicate the effect that surface texture might have on repeatability.

Finally, we can assess the effect of the measurement procedure on repeatability by making a series of separate measurements on the sample, repositioning the instrument each time, as we would if we were making the measurement at a later date. As this process includes all the factors that affect repeatability, it is often used to define a minimum significant colour difference. For example, Ford and Smith (2017) used CIE Lab values from ten measurements – between which the colorimeter was removed and repositioned – to calculate a statistical measurement of 'mean colour distance from the mean' (MCDM). Any colour difference less than double the sum of MCDM for the two readings was considered insignificant.

6.8.3 Reproducibility

As the preceding chapters have shown, most applications of colour measurement in the cultural heritage field focus on measuring colour differences to understand the nature of change, its rate and its effect on the appearance of objects. Whether we are studying objects or samples designed to help us understand the effects of ageing or treatment, we need to make readings over the course of time. If these measurements are to be comparable, we need to focus on their long-term reproducibility; we must be confident that a measured difference is due to a change in the object rather than in the instrument or procedure.

First, as our instruments may drift, we should recalibrate before a set of readings is taken. We generally assume that the standards that we use for calibration are themselves unchanging, but this can be confirmed, as necessary, by having them recalibrated by a specialist laboratory. Between longer-term measurements it is quite likely that the spectrometer or colorimeter will have been used for another purpose, so we need to ensure that its settings are returned to those used in earlier measurements – see, for example, the *Colorimetry checklist* in Chapter 3.

In practice, the largest single factor that affects the quality of data and the ability to make quantitative measurements in the longer term is the positional accuracy of remeasurement. One approach to minimise this potential issue is to study samples with high colour homogeneity and minimal texture, as the exact measurement position on the surface becomes less critical. This may not be possible for certain samples and is frequently impossible when studying object surfaces. Methods of relocating the instrument on the same point on an object or sample, such as by using a template or by imaging the area, are described in Section 6.3. Even with these aids to relocation, it can be important to document how exactly they will be used, particularly if the readings are taken over a long period, or another person may be responsible for future readings.

Some instances of poor reproducibility can be worked around. For example, although the graph of colour change against time in Figure 6.9 shows a rogue reading, the other readings appear to have been made in a consistent manner and this data point can be disregarded. Clearly, the more frequently data are collected, the easier it will be to isolate such anomalies. In contrast, the data in Figure 6.10 illustrate a less easily resolved situation where there is a clear discontinuity after 10 Mlux hours. This might be due to a new researcher applying a different procedure, or a new instrument being used that, even after calibration, differs fundamentally from that used originally.

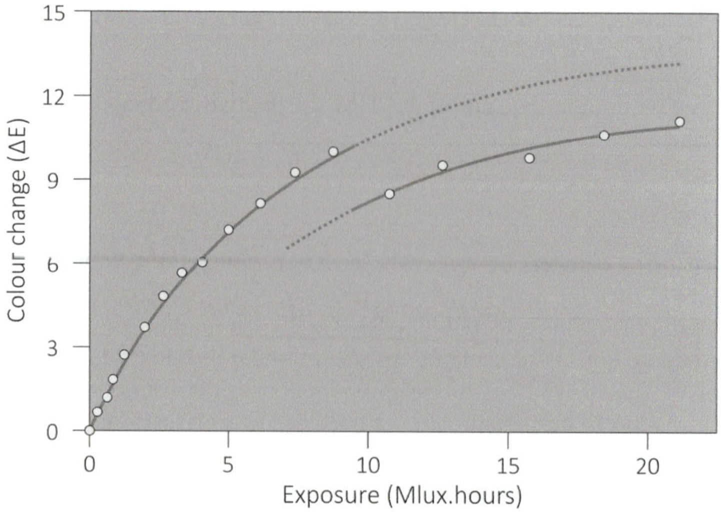

Figure 6.10 Plot of colour change against light exposure. A change of procedure or operator after 10 Mlux hours has produced a discontinuity in the readings

6.9 Presenting data

Before concluding this chapter, it is worth looking briefly at what colour data we might report and how they might be presented.

Spectral information in the visible region is generally presented graphically and we have seen many examples in the preceding chapters (e.g., Figures 2.17 and 6.7). The spectra should be accompanied by details of the instrument used and measurement parameters such as measurement geometry. For brevity, peak or edge wavelengths (in nm) are sometimes reported or tabulated, which can allow comparison between samples and reference standards without interrogating the spectra themselves. We have also seen that first derivative spectra are sometimes calculated to allow a peak or edge wavelength to be determined more confidently. In fluorimetry, the peak wavelength responsible for excitation ($\lambda_{excitation}$) and the peak wavelength for the emission ($\lambda_{emission}$) are often reported.

When colours are presented as images, we usually attempt to represent the colours as they appear to the eye, using standard colorimetric procedures, but bearing in mind the limitations of computer screens, data projectors or printers. Recent advances in imaging methods outside the visible region (e.g., hyperspectral imaging and XRF mapping) have been sufficiently rapid that no standards seem yet to have emerged for their presentation, with different researchers applying specific colour palettes to their visualisations (Spring et al. 2021).

Unless there are good reasons for doing otherwise, it is good practice to report colorimetric data in the CIE Lab system, which is the current standard. Again, details of the instrument and the measurement parameters given in the *Colorimetry checklist* in Chapter 3 should be recorded. If a spectrometer has been used to collect the data, it is worth giving details of the method by which the spectra were converted to colorimetric data – by the instrument or by the user, and, if the latter, using which method?

Since colorimetric measurements are often used to study alterations in colour, ΔE values are commonly used to express the magnitude of change and to give an indication of how visible the difference will appear. As we saw in Chapter 3, there is more than one method of calculating ΔE, each of which will produce a different value, so it is important to indicate which formula has been applied, either by the instrument or the user. It is much easier to calculate ΔE_{76}, but the current standard (ΔE_{00}) provides a better correspondence between the quantified colour change and its visibility. Perhaps because it is calculated automatically by many instruments, ΔE_{00} tends to be the norm in most current studies.

While ΔE provides an indication of the magnitude of colour change, used alone, it gives no information on the nature of that change. If the ΔE values are tabulated alongside the CIE Lab (or CIE LCh) data from which they are calculated (or component differences such as ΔL, Δa and Δb), it is usually possible to categorise the type of change more precisely, perhaps into one of the types of change described in Chapter 5. Even when space constraints preclude the publication of the full dataset, it is now often possible to include these in supplemental online data associated with the paper.

Colour change data are typically presented as a plot of ΔE against time or exposure (to light or other agents of change), but we have seen that it can be helpful also to graph the change in a single colour co-ordinate with time or pairs of colour co-ordinates. In this way it is possible to trace the yellowing of a polymer by plotting b* or Δb (or the Yellowness Index) against time. In an earlier example, we saw in Figure 5.11 the fading of a brazilwood lake as followed by plotting a* and b*, which revealed the generation of a yellow intermediate as the red lake pigment faded, something that would not be evident from the ΔE data alone and which was subsequently confirmed through the chemical identification of this intermediate (Peggie et al. 2018).

The number of significant figures to which the data are reported will depend on the accuracy and repeatability of measurement, but it is rare to find instances where it would be justified to report CIE Lab or ΔE values to more than two decimal places. There is also a tendency to place more weight on colour measurements than the instrumentation and procedure justify, with conclusions sometimes drawn from data that are at the level of experimental error.

In general, collecting more data can assist in improving the quality of results. It is common (and good) practice to make several readings on a sample and to average these results (some colorimeters allow successive readings to be taken until a predefined level of consistency has been obtained). Although not possible in a simple 'before and after' study, in some types of experiment, we can improve the quality of data by taking readings more frequently. The example presented earlier in Figure 6.9 shows how this can help to identify 'rogue' readings.

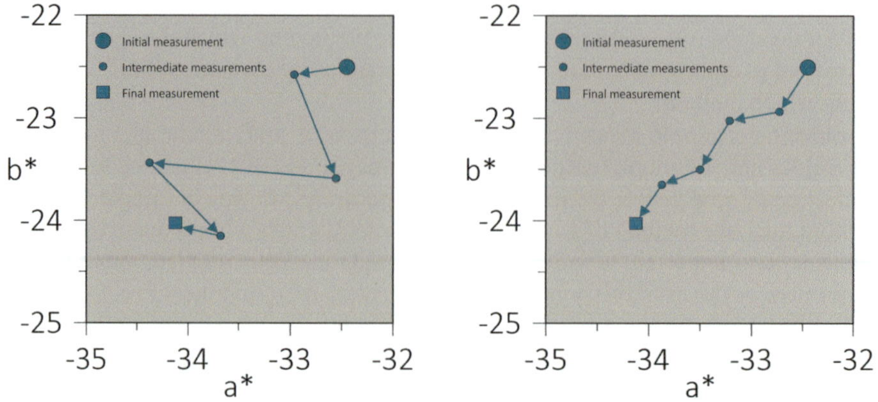

Figure 6.11 An example of how additional measurements can help inform the interpretation of colour changes, see text

Collecting more frequent readings can also allow us to interpret changes between successive data points. In Figure 6.11, we see two examples of five successive small changes in colour presented on a plot of a* versus b*. The overall change in both plots is relatively small and might have been dismissed as below experimental error if only the initial and final readings had been made. In the left plot, the successive changes in colour revealed by the intermediate readings have different directions and do not follow the overall direction of change, suggesting that they are at the level of experimental error. In contrast, the successive changes in colour in the right plot align well with the overall change, implying that although small, the overall change is meaningful.

7 Mitigating colour change

In cultural heritage institutions the care of collections relies on implementing preventive conservation measures that ensure good practices, processes and documentation. As preventive conservation is the planned topic of another book in this series, only those measures that help to mitigate changes in the colour of objects will be addressed here. Although visible and often dramatic, alterations in the colour of objects need to be seen in perspective of the more devastating effects of, for example, physical forces (earthquakes, mishandling, etc.), water (inundation and leaks) and fire. Changes in colour can, however, sometimes serve as a visible sign of reactions that might lead to more serious damage.

Preventive conservation uses a framework developed by the Canadian Conservation Institute (CCI) that defines ten **agents of deterioration** (see *Agents of deterioration* box) to help identify the risks to collections from different biological, physical, chemical and human sources (CCI 2017).

Ten agents of deterioration, as defined by the Canadian Conservation Institute (CCI)

- Physical forces
- Light, ultraviolet and infrared
- Water
- Incorrect relative humidity
- Incorrect temperature
- Pollutants
- Fire
- Pests
- Thieves and vandals
- Dissociation

This framework provides a useful method for us to consider the ways in which the external environment can interact with coloured materials. The CCI also

DOI: 10.4324/9781003397182-8

promotes a staged approach to controlling the effects of these agents: avoid, block, detect, respond and recover/treat. We will look primarily at how to avoid or block agents of deterioration in the sections that follow.

7.1 Physical forces

It is usually the structure rather than the colour of objects that we consider to be at risk from physical forces, but we have seen that some colours owe their colour to the microstructure of their surface and that this colour can be lost or altered by distortion or crushing (e.g., in some natural history specimens). These changes typically result from impacts, shocks or the exertion of pressure on an object. They can be avoided or blocked by ensuring objects are safely and adequately supported when on display and appropriately packed when transported or stored, avoiding, for example, stacking crushable objects on top of each other.

Physical changes to the surface from abrasion can change its gloss and scatter, producing a dull appearance, desaturating colour or – in extreme cases – obscuring underlying colours. Some abrasion results from mishandling, but dust accumulation and subsequent movement by air currents or during removal are a common source of micro-scratches. Tight-fitting or sealed boxes, display cases and glazed frames can lower dust accumulation when objects are in store or on exhibition. Avoiding dusty environments will reduce the issue, while careful removal – using vacuum cleaners rather than methods that rub the surface – minimises abrasion.

Vibration, particularly if present over a long period, may alter the surface structure, perhaps through the propagation of a crack network that alters the object's scattering properties. Steps are frequently taken to minimise vibration when building works occur in museums or during transportation, changing practices where possible or introducing cushioning materials. In extreme cases vibration (or impact and shock) can result in change through material loss, perhaps revealing a different underlying colour.

7.2 Light, ultraviolet and infrared

We have seen that radiation plays a critical role in colour change, often providing the energy necessary to instigate reactions that destroy or create chromophores. We can differentiate between visible light, which we need to see the form and colour of objects, and the regions beyond the visible, the ultraviolet and infrared, which we may want to exclude to reduce deterioration. Of these two invisible regions, ultraviolet radiation is of most concern since, as we saw in Chapter 1, it has higher energy and is thus more likely instigate change. Lower-energy infrared radiation is less damaging, although its removal potentially helps to reduce heat gain and, therefore, limit reaction rates.

To minimise ultraviolet light exposure, we can employ light sources that have a low ultraviolet output, including most LED and some fluorescent lamps. Other light sources, including daylight, some fluorescent lamps and tungsten-halogen lamps (which are becoming obsolete), emit higher levels of ultraviolet, but filters can be used to block most or all such radiation. Modern windows and rooflights often contain a laminating interlayer that absorbs ultraviolet and older glazing can be treated with films or varnishes that fulfil this function. Contrary to some guidance, the removal of ultraviolet will not eliminate radiation-induced damage, as visible light can also instigate some reactions that lead to colour change. However, as light is needed to see objects, reducing deterioration must be balanced with providing sufficient light for good access to objects for appreciation and study.

Since light damage is generally proportional to light dose, which is the product of light level and length of exposure (see Chapter 6, *Reciprocity* box), strategies to reduce deterioration address either or both factors. We are somewhat limited by how much we can lower light levels if we wish to see objects well, and seeing them well will require more light for some (e.g., older) visitors than others. Strategies for moderating light levels include making sure our visual system is well adapted and avoiding distracting glare and reflections. More opportunities exist to reduce the length of exposure; we can place objects in dark storage when they are not on view and restrict the time they are lit when on view by using sensors to turn lights off when there are no visitors. We can also ensure that spaces are kept in darkness outside opening hours, except when light is needed for events, security or cleaning.

As we saw in Chapter 6, some materials are more sensitive to light than others, so we can determine – from microfade testing or studies on surrogate materials – those objects that are most likely to be at risk and focus our preventive conservation measures on reducing their exposure to light. It is common practice to set a budget for exposure over a particular period – often five or ten years – and to use this to limit periods on display. In addition to vulnerability, we might also want to consider value or significance when setting such light budgets. More detail on the effect of light on collections and light management can be found in *Museum Lighting* (Saunders 2020).

7.3 Water and incorrect relative humidity

When we list 'water' as an agent of deterioration, we are generally drawing attention to the possibility of inundation of collections through flooding, failure of the building fabric during wet weather or leaking water pipes. The resulting saturation with liquid water can lead to the dissolution and movement of colorants within objects (bleed), transport soluble materials to the surface or provide an environment in which biological and chemical activity is more likely. Each may increase the possibility of changes to colours, as in the earlier example of the reaction between copper-containing pigments and chloride ions in wall paintings (see Figure 4.11). It is worth noting, however, that under these circumstances changes in colour are likely to be of less concern than more severe structural effects from prolonged contact with water.

In addition to liquid water, we may see damage to colours from the high concentration of water vapour that is present at elevated relative humidity (RH), which can also act as a solvent to facilitate the reaction between colorants and other materials, for example, by dissolving oxygen, carbon dioxide or pollutant gases. In previous sections we also saw that hydrolysis plays a role in many degradation processes. Because water or hydroxyl ions are involved in such reactions, rates of hydrolytic degradation can be lowered by maintaining a drier environment.

Studies have shown that at higher RHs the presence of moisture also causes the rates of other reactions to increase, including those triggered by light exposure (Kühn 1968; Saunders and Kirby 2004). This could be a result of the solubilisation of other components – for example, acidic materials in a binding medium, textile or paper substrate, or atmospheric pollutants – that are key to the degradation of the colorant.

Although the point at which the RH is considered 'incorrect' will vary depending on the reaction we are considering, we can define critical boundary RHs for some common reactions. As we saw earlier, metals, particularly archaeological iron and bronze, are prone to corrode in the presence of moisture. A complex sequence of critical RH values, each due to a specific compound in the chain of corrosion, determines the rate of deterioration. Generally, the lower the RH in which metals are kept, the better, with desiccated storage and display quite commonly applied. A critical RH for copper alloys in the 35–40% region has been proposed by Thickett (2018), while storage at less than 40–50% has been shown to reduce corrosion in iron (Watkinson, Rimmer and Emmerson 2019).

At high RH, a critical upper boundary of >70–75% is based on the need to avoid mould germination, but will also reduce the rates of hydrolytic and other reactions that proceed rapidly at higher RH, for example, the deterioration of cellulose acetate and nitrate films.

Certain materials absorb and desorb moisture in response to changes in RH. As a result, the prevailing RH for these hygroscopic materials is often kept within a particular range. This avoids an uptake or loss of moisture that will lead to swelling or contraction that can trigger structural changes in, for example, wood, textiles or paper. Structural damage is often most severe if the RH cycles periodically, leading to sequential changes in dimensions that set up stresses in the object that may lead to cracking, delamination and ultimately failure. As we saw in Chapter 5, the development of craquelure or the delamination of layers may alter the colour of objects as they affect both surface gloss and the balance between scatter and absorption within materials.

In addition to these general boundaries for RH, there are some specific materials that are affected at a particular RH. Certain unstable glasses need a narrow range of RH as one of their components may deliquesce (absorb water into its chemical structure) above a certain point, causing the phenomenon of 'weeping glass', while at lower RH other components dehydrate, creating a fine network of cracks, termed crizzling (Koob et al. 2018). Some mineral specimens have specific critical RHs above or below which they can crumble and/or weep (Waller 1992).

The RH that is appropriate for any particular object will depend on historical averages in the region where it has been acclimatised over decades or centuries, but keeping conditions within the critical upper and lower boundaries for mould growth and desiccation respectively is advised. Within these limits, lower RHs will tend to reduce reaction rates; a rule of thumb for less stable materials – akin to that expressed for temperature by the Arrhenius relationship (see Chapter 4, Equation 4.1) – suggests that the rate may be more than halved if the RH is halved (Michalski 2002).

In recent years, guidelines for the control of RH have recommended broader bands, recognising that maintaining very tight conditions can be costly, energy-intensive and counter to wider sustainability goals. Where narrower ranges of RH are needed for preservation of specific objects, microclimate display and storage enclosures have become more common, avoiding the need to maintain a particular climate in a whole room or building.

7.4 Incorrect temperature

Low or high temperatures can cause structural changes to objects. At low temperatures some materials become brittle and more prone to damage by shock or vibration while at higher temperatures other materials melt or soften, which may lead to irreversible deformation. Temperature changes can cause expansion or contraction, which could theoretically alter structural colours, but this is not a commonly reported phenomenon.

The main influence of temperature on colour change is an increase in reaction rate as the temperature rises, which we covered in Chapter 4 when we considered the Arrhenius relationship (see Equation 4.1) and Michalski's rule of thumb that reactions in organic materials double in rate for a rise of around 5°C.

The assumption is, therefore, that we should keep temperatures as low as possible to reduce chemical reaction rates. This is the principle behind cold storage in archive collections and for very reactive materials such as acetate or nitrate films. Such cold temperatures are not suitable for display or study environments, or in stores that are in frequent use, where the temperature is usually set for visitor or staff comfort. In temperate climates, this comfort zone is generally considered to fall between 18° and 27°C, depending on the season.

There are a few examples of materials that demonstrate thermochromism, a reversible change of colour with temperature, but these are not widespread in the cultural heritage sector and most involve the loss of water at higher temperatures and its reabsorption as the material cools.

7.5 Pollutants

Generally, liquid and solid pollutants tend to interfere with the colour of objects through changes to the physical properties of the surface. Colourless volatile liquids can leave a 'tide mark' as they evaporate, while coloured liquids (red wine is a frequently quoted example) can stain porous materials. Higher boiling-point liquids, such as oils or greases, can penetrate porous materials altering their transparency, or coat the surface, changing its gloss and – because they are often sticky – increasing the retention of dust.

Liquid spillages are not common – at least in museums that discourage eating and drinking in display spaces – but grease accumulation and impregnation may be an issue when porous materials such as sculptures, ceramics or textiles can be touched while on display. Where objects are kept out of reach, touchable surrogate materials are sometimes provided to allow visitors to appreciate the feel of the original.

Solid pollutants come from a variety of sources. Although many cities now have much less polluted atmospheres than when coal and wood were common fuels, particulate matter can still settle on surfaces – and some objects still bear dirt from that period. Dust can also originate from building materials, such as concrete, or be shed by visitors' clothing, skin or hair. Visitors are also a vector for the transport of dust into buildings from outside sources.

Dust accumulation can obscure underlying colours, reduce contrast between colours and thus alter the appearance of objects, making their appreciation and interpretation more difficult. Some dust can induce chemical changes if left *in situ* and – as we saw in Section 7.1 – dust removal must be effected carefully to avoid physical damage to the surface. Periodic removal can be an important preventive measure, ensuring that dust does bind more strongly to an object's surface or become engrained in its texture. Where dirt is not ingrained, it is usually removed by controlled cleaning; the use of a vacuum cleaner fitted with a high-efficiency particulate air (HEPA) filter that can remove very fine particles is often recommended.

In most display and storage spaces some dust is inevitable, but measures can be taken to reduce levels. Good seals on doors and windows and self-closing or revolving external doors can reduce dust ingress while using druggets (protective carpeting, or a transparent material printed to replicate the flooring it is shielding) at external doors and walking routes can lower dust penetration on visitors' feet. When building work is undertaken within the museum or house, effective seals between 'clean' and 'dirty' zones help to reduce dust accumulation on displays. Placing objects in enclosures (e.g., boxes or folders) for storage or cases for display can help but the latter may not be compatible with the ethos of, for example, a historic house display (Lloyd et al. 2007).

7.6 Oxygen-free atmospheres

Oxygen does not feature in the list of agents of deterioration and yet, as we have seen, it is implicated in many of the reactions that cause colour change – and indeed in deteriorative processes that affect the structural integrity of the organic materials we find in many cultural heritage objects. There are several scientific studies indicating that many colour changes observed in such collections are – often dramatically – reduced by applying anoxic conditions, although caution is needed, as some pigments, such as Prussian blue and iron gall inks are known to fare better in the presence of some oxygen (Beltran, Druzik and Maekawa 2012; Lerwill et al. 2015).

Despite this evidence, there are comparatively few examples of the use of anoxic storage or display environments, principally for pragmatic reasons. As we need oxygen to breathe, it is impractical to maintain low-oxygen display or storage areas, with examples limited to automated storage facilities from which objects are removed to 'inhabited' parts of the building by robots.

There is a longer history of enclosing individual sensitive objects in oxygen-free storage or display environments. For example, anoxic conditions have been found to be beneficial for the storage of plastics and rubber (Dyer et al. 2011). Towards the end of the nineteenth century a Turner painting was placed in a passive display case from which the air had been removed (Brommelle 1964). More recently, the US Library of Congress has developed actively maintained low-oxygen cases for some of its key original documents, and oxygen-free environments have been provided for the royal mummies displayed in the Cairo Museum (Maekawa 1998), now in the National Museum of Egyptian Civilization.

Both low-oxygen stores and enclosures require maintenance and monitoring as – even in the best systems – oxygen will gradually leak or diffuse into the space; Brommelle (1964) speculated that the improved preservation of the Turner painting mentioned above was due more to its encasement (and inaccessibility for cycles of conservation treatment) than the absence of oxygen.

References

Ambers, J., C. Higgitt and D. Saunders (eds). 2010. *Italian Renaissance Drawings: Technical Examination and Analysis*. London: Archetype and British Museum.

ASTM. 2005. *Standard Practice for Calculating Yellowness and Whiteness Indices from Instrumentally Measured Color Coordinates. ASTM E313-05*. West Conshohocken, PA: ASTM International.

ASTM. 2010. *Standard Test Methods for Lightfastness of Colorants Used in Artists' Materials. ASTM D4303-10*. West Conshohocken, PA: ASTM International.

ASTM. 2016. *Standard Test Methods for Measuring the Hiding Power of Powder Coatings. ASTM D6441-05*. West Conshohocken, PA: ASTM International.

Aze, S. and J.-M. Vallet. 2002. "Chromatic Degradation Processes of Red Lead Pigment". In *ICOM Committee for Conservation, 13th Triennial Meeting, Rio de Janeiro, 20–27 September 2002*, ed. R. Vontobel, 549–555. London: James and James.

Baechler, G., A. Latty, M. Pacholska, M. Vetterli and A. Scholefield. 2021. "Shedding Light on 19th Century Spectra by Analyzing Lippmann Photography". *Proceedings of the National Academy of Sciences* 118: e2008819118.

Bailie, C. W., R. M. Johnston-Feller and R. L. Feller. 1988. "The Fading of Some Traditional Pigments as a Function of Relative Humidity". In *Materials Issues in Art and Archaeology: Materials Research Society Symposium Proceedings, no. 123*, eds E.V. Sayre, P. Vandiver, J. Druzik and C. Stevenson, 287–292. Warrendale, PA: Materials Research Society.

Beltran, V. L., J. Druzik and S. Maekawa. 2012. "Large-Scale Assessment of Light-Induced Colour Change in Air and Anoxic Environments". *Studies in Conservation* 57: 42–57.

Beltran, V. L., C. Pesme, S. K. Freeman and M. Benson. 2021. *Microfading Tester: Light Sensitivity Assessment and Role in Lighting Policy*. Los Angeles, CA: Getty Conservation Institute.

Berns, R. S. 2016. *Color Science and the Visual Arts: A Guide for Conservators, Curators, and the Curious*. Los Angeles, CA: Getty Conservation Institute.

Berns, R. S. 2019. "Digital Color Reconstructions of Cultural Heritage Using Color-Managed Imaging and Small-Aperture Spectrophotometry". *Color Research and Technology* 44: 531–546.

Berns, R. S., F. W. Billmeyer and M. Saltzman. 2000. *Billmeyer and Saltzman's Principles of Color Technology*. 3rd edn. New York: Wiley.

Berns, R. S., F. H. Imai, L. A. Taplin, S. Byrns, F. Casadio, I. Fiedler, C. Gallagher and A. Newman. 2005. "Rejuvenating the Appearance of Seurat's *A Sunday on La Grande Jatte*

1884 Using Colour and Imaging Science Techniques: A Simulation". In *ICOM Committee for Conservation, 14th Triennial Meeting, The Hague, 12–16 September 2005*, ed. I. Verger, 452–458. London: James & James/Earthscan.

Berrie, B. 2007. *Artists' Pigments: A Handbook of Their History and Characteristics*. Vol. 4. Washington, DC: National Gallery of Art.

Berrie, B. H. and S. Q. Lomax. 1997. "Azo Pigments: Their History, Synthesis, Properties, and Use in Artists' Materials". In *Conservation Research 1996/1997. Studies in the History of Art 57*, 9–33. Washington, DC: National Gallery of Art.

Bescoby, J., J. Rayner and S. Tanimoto. 2010. "Dry Drawing Media". In *Italian Renaissance Drawings: Technical Examination and Analysis*, eds J. Ambers, C. Higgitt and D. Saunders. London: Archetype and British Museum.

Bomford, D., J. Kirby, J. Leighton and A. Roy. 1990. *Art in the Making: Impressionism*. London: National Gallery.

Boyce, P. R. 2014. *Human Factors in Lighting*. 3rd edn. Boca Raton, FL: CRC Press.

Broers, F. T. H., K. Janssens, J. Nelson Weker, S. M. Webb, A. Mehta, F. Meirer and K. Keune. 2023. "Two Pathways for the Degradation of Orpiment Pigment (As_2S_3) Found in Paintings". *Journal of the American Chemical Society* 145: 8847–8859.

Brommelle, N. S. 1964. "Russell and Abney Report on the Action of Light on Water Colours". *Studies in Conservation* 9(4): 140–152.

Bullock, L. 1978. "Reflectance Spectrophotometry for Measurement of Colour Change". *National Gallery Technical Bulletin* 2: 48–55.

CAMEO (*Conservation & Art Materials Encyclopedia* Online). n.d. Available at: https://cameo.mfa.org/ (accessed 16 August 2024).

Cardon, D. 2007. *Natural Dyes: Sources, Tradition, Technology and Science*. London: Archetype.

Carlyle, L. and J. H. Townsend. 1990. "An Investigation of Lead Sulphide Darkening of Nineteenth Century Painting Materials". In *Dirt and Pictures Separated*, ed. V. Todd, 40–43. London: UKIC.

CCI (Canadian Conservation Institute). 2017. "Agents of Deterioration". Canadian Conservation Institute. Available at: www.canada.ca/en/conservation-institute/services/agents-deterioration.html (accessed 16 August 2024).

Chevreul, M. E. 1864. *Des couleurs et de leurs applications aux arts industriels à l'aide des cercles chromatiques*. Paris: J. B. Ballière et fils.

Christie, R. M. 2015. *Colour Chemistry*. 2nd edn. Cambridge: Royal Society of Chemistry.

CIE (Commission Internationale de l'Éclairage). 1978. *Recommendations on Uniform Color Spaces, Color-Difference Equations, Psychometric Color Terms. Publication no. 15, sup. 2*. Paris: Bureau central de la CIE.

CIE (Commission Internationale de l'Éclairage). 1995a. *Industrial Colour-Difference Evaluation. CIE Technical Report 116-1995*. Vienna: CIE Central Bureau.

CIE (Commission Internationale de l'Éclairage). 1995b. *Method of Measuring and Specifying Colour Rendering Properties of Light Sources. Publication CIE, no. 13.3-1995*. Vienna: CIE Central Bureau.

CIE (Commission Internationale de l'Éclairage). 2001. *Improvement to Industrial Colour Difference Evaluation. CIE Technical Report 142-2001*. Vienna: CIE Central Bureau.

CIE (Commission Internationale de l'Éclairage). 2004a. *Colorimetry*. 3rd edn. *CIE Technical Report 15*. Vienna: Commission Internationale de l'Éclairage.

CIE (Commission Internationale de l'Éclairage). 2004b. *Colour Appearance Model for Colour Management Systems: CIECAM02. CIE Technical Report 159:2004*. Vienna: CIE Central Bureau.

CIE (Commission Internationale de l'Éclairage). 2009. *Indoor Daylight Illuminants. CIE Technical Report 184:2009.*Vienna: CIE Central Bureau.

CIE (Commission Internationale de l'Éclairage). 2010. *Practical Daylight Sources for Colorimetry. CIE Technical Report 192:2010.* Vienna: CIE Central Bureau.

CIE (Commission Internationale de l'Éclairage). 2017. *Colour Fidelity Index for Accurate Scientific Use. CIE Technical Report 224:2017.* Vienna: CIE Central Bureau.

CIE (Commission Internationale de l'Éclairage). 2022a. *Colorimetry — Part 2: CIE Standard Illuminants. CIE Technical Report 11664-2:2022.* Vienna: CIE Central Bureau.

CIE (Commission Internationale de l'Éclairage). 2022b. *Colorimetry — Part 6: CIEDE2000 Colour-Difference Formula. CIE Technical Report 11664-6:2022.* Vienna: CIE Central Bureau.

CIE (Commission Internationale de l'Éclairage). 2022c. *The CIE 2016 Colour Appearance Model for Colour Management Systems: CIECAM16. CIE Technical Report 248:2022.* Vienna: CIE Central Bureau.

CIE (Commission Internationale de l'Éclairage). n.d. *Datasets*. Available at: https://cie.co.at/data-tables (accessed 16 August 2024).

Clarke, F. J. J., R. McDonald and B. Rigg. 1984. "Modification to the JPC79 Colour-Difference Formula". *Journal of the Society of Dyers and Colourists* 100(4): 128–132.

Cosentino, A. 2014. "FORS Spectral Database of Historical Pigments in Different Binders". *e-conservation Journal* 2: 57–68.

Daniels, V., and B. Leach. 2004. "The Occurrence and Alteration of Realgar on Ancient Egyptian Papyri". *Studies in Conservation* 49(2): 73–84.

Dei, L., A. Ahle, P. Baglioni, D. Dini and E. Ferroni. 1998. "Green Degradation Products of Azurite in Wall Paintings: Identification and Conservation Treatment". *Studies in Conservation* 43: 80–88.

de Keijzer, M. 2014. "The Delight of Modern Organic Pigment Creations". In *Issues in Contemporary Oil Paint*, eds K. J. van den Berg, A. Burnstock, M. de Keijzer, J. Krueger, T. Learner, A. de Tagle and G. Heydenreich, 45–73. Cham: Springer.

Delaney, J. K., D. M. Conover, K. A. Dooley, L. Glinsman, K. Janssens and M. Loew. 2018. "Integrated X-Ray Fluorescence and Diffuse Visible-To-Near-Infrared Reflectance Scanner for Standoff Elemental and Molecular Spectroscopic Imaging of Paints and Works on Paper". *Heritage Science* 6: 31.

Delaney, J. K. and K. A. Dooley. 2022. "Visible and Infrared Reflectance Imaging Spectroscopy of Paintings and Works on Paper". In *Analytical Chemistry for the Study of Paintings and the Detection of Forgeries*, eds M. P. Colombini, I. Degano and A. Nevin, 115–132. Cham: Springer.

de la Rie, E. R. and C. W. McGlinchey. 1990. "The Effect of a Hindered Amine Light Stabilizer on the Aging of Dammar and Mastic Varnish in an Environment Free of Ultraviolet Light". *Studies in Conservation* 35(suppl. 1): 160–164.

Dichter, D. W. 2023. "Kubelka–Munk Model of Full-Gamut Oil Colour Mixing". *Journal of the International Colour Association* 32: 70–78.

Dietemann, P., C. Higgitt, M. Kälin, M. J. Edelmann, R. Knochenmuss and R. Zenobi. 2009. "Aging and Yellowing of Triterpenoid Resin Varnishes: Influence of Aging Conditions and Resin Composition". *Journal of Cultural Heritage* 10(1): 30–40.

Dooley, K. A., A. Chieli, A. Romani, S. Legrand, C. Miliani, K. Janssens and J. K. Delaney. 2020. "Molecular Fluorescence Imaging Spectroscopy for Mapping Low Concentrations of Red Lake Pigments: Van Gogh's Painting *The Olive Orchard*". *Angewandte Chemie, International Edition* 59(15): 6046–6053.

Dooley, K. A., J. Coddington, J. Krueger, D. M. Conover, M. Loew and J. K. Delaney. 2017. "Standoff Chemical Imaging Finds Evidence for Jackson Pollock's Selective Use of Alkyd and Oil Binding Media in a Famous 'Drip' Painting". *Analytical Methods* 9: 28–37.

Down, J. L. 2001. "Review of CCI Research on Epoxy Resin Adhesives for Glass Conservation". *Studies in Conservation* 46(suppl. 1): 39–46.

Dyer, J., C. Ward, N. Rode, M. Hacke and Y. Shashoua. 2011. "Reassessment of Anoxic Storage of Ethnographic Rubber Objects in the British Museum Collections". In *ICOM Committee for Conservation, 16th Triennial Meeting, Lisbon, 19–23 September 2011*, ed. J. Bridgland. Lisbon: Critério—Produção Grafica.

Eastaugh, N., V. Walsh, T. Chaplin and R. Siddall. 2004. *The Pigment Compendium: A Dictionary and Optical Microscopy of Historical Pigments*. Amsterdam: Butterworth-Heinemann.

Eisen, E. X. 2023. "Beetle Carapaces in Basohli Miniature Paintings (ca. 1660–1700)". *The Public Domain Review*. Available at: https://publicdomainreview.org/collection/basohli-beetle-paintings/ (accessed 16 August 2024).

Eng, C. W., F. D. Preusser and T. T. Schaeffer. 2016. "Reflections on Light Monitoring: Evaluating Museum Lighting Options for Modern and Contemporary Art". *Studies in Conservation* 61 (suppl. 2): 44–48.

Fairchild, M. D. 2005. *Color Appearance Models*. Wiley-IS&T Series in Imaging Science and Technology. 2nd edn. Chichester: Wiley.

Feller, R. L. 1975. "Studies on Photochemical Deterioration". In *ICOM Committee for Conservation, 4th Triennial Meeting, Venice, 13–18 October 1975*, 75/19/1–8. Paris: International Council of Museums.

Feller, R. L. 1986. *Artists' Pigments: A Handbook of Their History and Characteristics*. Vol. 1. Washington, DC: National Gallery of Art.

Ferguson, J. 2014. "Munsell Notations and Color Names: Recommendations for Archaeological Practice". *Journal of Field Archaeology* 39(4): 327–335.

Fonseca, B., C. Schmidt Patterson, M. Ganio, D. MacLennan and K. Trentelman. 2019. "Seeing Red: Towards an Improved Protocol for the Identification of Madder- and Cochineal-Based Pigments by Fiber Optics Reflectance Spectroscopy (FORS)". *Heritage Science* 7: 92.

Ford, B. L. 1992. "Monitoring Colour Change in Textiles on Display". *Studies in Conservation* 37(1): 1–11.

Ford, B. and N. Smith. 2010. "Protecting the Most Important, Most Exhibited and Most Fugitive Museum Objects from Light-Fading". *The Textile Specialty Group Postprints* 20: 157–186.

Ford, B. and N. Smith. 2017. "A Reality Check for Microfade Testing: Five Examples". In *ICOM Committee for Conservation, 18th Triennial Meeting Copenhagen 4–8 September 2017*, ed. J. Bridgland, art. 1508. Paris: International Council of Museums.

Giovannoni, S., M. Matteini and A. Moles. 1990. "Studies and Developments Concerning the Problem of Altered Lead Pigments in Wall Painting". *Studies in Conservation* 35(1): 21–25.

Gutscher, D., B. Mühlethaler, A. Portmann and A. Reller. 1989. "Conversion of Azurite into Tenorite". *Studies in Conservation*, 34(3): 117–122.

Hård, A., L. Sivik and G. Tonnquist. 1996a. "NCS, Natural Color System—From Concept to Research and Applications. Part I". *Color Research and Application* 21(3): 180–205.

Hård, A., L. Sivik and G. Tonnquist. 1996b. "NCS, Natural Color System—From Concept to Research and Applications. Part II". *Color Research and Application* 21(3): 206–20.

Healy, T. n.d. "Spectral Measurements of 'Neon' Lights in Artworks for the Identification of Gas Composition and Color Characterization". Available at: www.taylorhealy.com/neon (accessed 16 August 2024).

Hendriks, E., A. W. Brokerhof and K. van den Meiracker. 2017. "Valuing Van Gogh's Colours: From the Past to the Future". In *ICOM Committee for Conservation, 18th Triennial Meeting, Copenhagen, 4–8 September 2017*, ed. J. Bridgland, art. 1306. Paris: International Council of Museums.

Huang, S., D. Messinger and D. Bove. 2020. "Hyperspectral Analysis: Black Inks Separation in the Gough Map of Great Britain". *Proceedings of the SPIE 11392: Algorithms, Technologies, and Applications for Multispectral and Hyperspectral Imagery* XXVI. 113920.

Hunt, R. W. G. and M. Pointer. 2011. *Measuring Colour*. 4th edn. Wiley-IS&T Series in Imaging Science and Technology. Chichester: Wiley.

IESNA (Illuminating Engineering Society of North America). 2015. *IES Method for Evaluating Light Source Color Rendition. IES TM-30-15*. New York: IESNA.

IESNA (Illuminating Engineering Society of North America). 2018. *IES Method for Evaluating Light Source Color Rendition. IES TM-30-18*. New York: IESNA.

IFAC (Istituto di Fisica Applicata "Nello Carrara"). 2021. "Fiber Optics Reflectance Spectra (FORS) of Pictorial Materials in the 270–1700 nm Range". Available at: https://spectradb.ifac.cnr.it/fors/ (accessed 16 August 2024).

Imai, F. H., R. S. Berns and D.-Y. Tzeng. 2000. "A Comparative Analysis of Spectral Reflectance Estimated in Various Spaces Using a Trichromatic Camera System". *Journal of Imaging Science and Technology* 44(4): 280–287.

ISO (International Organization for Standardization). 2014a. *Textiles—Tests for Colour Fastness. Part B01, Colour Fastness to Light: Daylight. ISO 105-B01:2014*. Geneva: ISO.

ISO (International Organization for Standardization). 2014b. *Textiles—Tests for Colour Fastness. Part B02, Colour Fastness to Artificial Light: Xenon Arc Fading Lamp Test. ISO 105-B02:2014*. Geneva: ISO.

Izzo, F. C., M. Kratter, A. Nevin and E. Zendri. 2021. "A Critical Review on the Analysis of Metal Soaps in Oil Paintings". *Chemistry Open* 10(9): 904–921.

Jana, J., M. Ganguly and T. Pal. 2016. "Enlightening Surface Plasmon Resonance Effect of Metal Nanoparticles for Practical Spectroscopic Application". *RSC Advances* 6(89): 86174–86211.

Johnston-Feller, R. 2001. *Color Science in the Examination of Museum Objects: Non-Destructive Procedures. Tools for Conservation*. Los Angeles, CA: Getty Conservation Institute. Available at: http://hdl.handle.net/10020/gci_pubs/color_science

Johnston-Feller, R. M., R. L. Feller, C. W. Bailie and M. Curran. 1984. "The Kinetics of Fading: Opaque Paint Films Pigmented with Alizarin Lake and Titanium Dioxide". *Journal of the American Institute for Conservation* 23: 114–129.

Keune, K., J. Mass, F. Meirer, C. Pottasch, A. van Loon, A. Hull, J. Church, E. Pouyet, M. Cottegh and A. Mehtai. 2015. "Tracking the Transformation and Transport of Arsenic Sulfide Pigments in Paints: Synchrotron-Based X-Ray Micro-Analyses". *Journal of Analytical Atomic Spectrometry* 30: 813–827.

Kirby, J. 1977. "A Spectrophotometric Method for the Identification of Lake Pigment Dyestuffs". *National Gallery Technical Bulletin* 1: 35–45.

Kirby, J. 1999. "The Painter's Trade in the Seventeenth Century: Theory and Practice". *National Gallery Technical Bulletin* 20: 5–49.

Kirby, J. 2017. *The Diversity of Dyes in History and Archaeology*. London: Archetype.

Kirby, J. and D. Saunders. 1998. "Sixteenth- to Eighteenth-Century Green Colours in Landscape and Flower Paintings: Composition and Deterioration". *Studies in Conservation* 43(suppl. 1): 155–159.

Kirby, J. and D. Saunders. 2004. "Fading and Colour Change of Prussian Blue: Methods of Manufacture and the Influence of Extenders". *National Gallery Technical Bulletin* 25: 73–99.

Kirby, J., M. Spring and C. Higgitt. 2005. "The Technology of Red Lake Pigment Manufacture: Study of the Dyestuff Substrate". *National Gallery Technical Bulletin* 26: 71–87.

Kirby, J., M. van Bommel and A. Verhecken (eds). 2014. *Natural Colorants for Dyeing and Lake Pigments: Practical Recipes and Their Historical Sources*. London: Archetype.

Knoblauch, K., F. Saunders, M. Kudusa, R. Hynes, M. Podgor, K. E. Higgins and F. M. de Monasterio. 1987. "Age and Illuminance Effects in the Farnsworth-Munsell 100-Hue Test". *Applied Optics* 26(8): 1441–1448.

Kolar, J. and M. Strlič (eds). 2006. *Iron Gall Inks: On Manufacture, Characterisation, Degradation and Stabilisation*. Ljubljana: National and University Library.

Koob, S. P., N. A. R. van Giffen, J. J. Kunicki-Goldfinger and R. H. Brill. 2018. "Caring for Glass Collections: The Importance of Maintaining Environmental Controls". *Studies in Conservation* 63(suppl. 1): 146–150.

Kubelka, P. and F. Munk. 1931. "Ein Beitrag zur Optik der Farbanstriche". *Zeitschrift für Technische Physik* 12: 593–601.

Kuehni, R. G. and A. Schwarz. 2008. *Color Ordered: A Survey of Color Systems from Antiquity to the Present*. Oxford: Oxford University Press.

Kühn, H. 1968. "The Effect of Oxygen, Relative Humidity and Temperature on the Fading-Rate of Watercolors. Reduced Light-Damage in Nitrogen Atmosphere". *Studies in Conservation* 12 (suppl. 1): 79–88.

Kuzio, O. and S. Farnand. 2022. "Comparing Practical Spectral Imaging Methods for Cultural Heritage Studio Photography". *Journal on Computing and Cultural Heritage* 16(1): article 11.

Lafontaine, R. H. 1986. "Seeing through a Yellow Varnish: A Compensating Illumination System". *Studies in Conservation* 31(3): 97–102.

Lerwill, A., J. H. Townsend, J. Thomas, S. Hackney, C. Caspers and H. Liang. 2015. "Photochemical Colour Change for Traditional Watercolour Pigments in Low Oxygen Levels". *Studies in Conservation* 60(1): 15–32.

Liang, H., D. Saunders and J. Cupitt. 2005. "A New Multispectral Imaging System for Examining Paintings". *Journal of Imaging Science and Technology* 49(6): 551–562.

Lloyd, H., C. Bendix, P. Brimblecombe and D. Thickett. 2007. "Dust in Historic Libraries". In *Museum Microclimates*, eds T. Padfield and K. Borchersen, 135–144. Copenhagen: National Museum of Denmark.

Louis, C. 2017. "Gold Nanoparticles in the Past: Before the Nanotechnology Era". In *Gold Nanoparticles for Physics, Chemistry and Biology*, eds C. Louis and O. Pluchery, 2nd edn. 1–28. London: Imperial College.

Lussier, S. M. and G. D. Smith. 2007. "A Review of the Phenomenon of Lead White Darkening and its Conversion Treatment". *Studies in Conservation* 52(suppl. 1): 41–53.

Maekawa, S. (ed.). 1998. *Oxygen-Free Museum Cases. Research in Conservation*. Los Angeles, CA: Getty Conservation Institute. Available at: http://hdl.handle.net/10020/gci_pubs/oxygen_free_cases (accessed 16 August 2024).

Malkin, F., J. A. Larkin, J. F. Verrill and R. H. Wardman. 1997. "The BCRA–NPL Ceramic Colour Standards, Series II – Master Spectral Reflectance and Thermochromism Data". *Journal of the Society of Dyers and Colourists* 113: 84–94.

Matteini, M. and A. Moles. 1981. "The Reconversion of Oxidized White Lead in Mural Paintings: A Control after a Five Year Period". In *ICOM Committee for Conservation, 7th Triennial Meeting, Ottawa, 21–25 September 1981*, ed. R. Vontobel, 81/15/1–8. Paris: International Council of Museums.

McFarland, M. R. 1997. "The Whitening Effects of Peroxide Gels on Darkened Lead White Paint". In *The Book and Paper Group Annual* vol. 16: 55–65. Washington, DC: The American Institute for Conservation.

Michalski, S. 2002. "Double the Life for Each Five-Degree Drop, More Than Double the Life for Each Halving of Relative Humidity". In *ICOM Committee for Conservation, 13th Triennial Meeting, Rio de Janeiro, 20–27 September 2002*, ed. R. Vontobel, 66–72. London: James and James.

Michalski, S. 2018. "Agent of Deterioration: Light, Ultraviolet and Infrared". Canadian Conservation Institute. Available at: www.canada.ca/en/conservation-institute/services/agents-deterioration/light.html (accessed 16 August 2024).

Modugno, F., F. Di Gianvincenzo, I. Degano, I. D. van der Werf, I. Bonaduce and K. J. van den Berg. 2019. "On the Influence of Relative Humidity on the Oxidation and Hydrolysis of Fresh and Aged Oil Paints". *Scientific Reports* 9: 5533.

Monico, L., K. Janssens, E. Hendriks, F. Vanmeert, G. Van der Snickt, M. Cotte, G. Falkenberg, B. G. Brunetti and C. Miliani. 2015. "Evidence for Degradation of the Chrome Yellows in Van Gogh's *Sunflowers*: A Study Using Noninvasive *in situ* Methods and Synchrotron-Radiation-Based X-Ray Techniques". *Angewandte Chemie, International Edition* 54: 13923–13927.

Monico, L., F. Rosi, R. Vivani, L. Cartechini, K. Janssens, N. Gauquelin, D. Chezganov, J. Verbeeck, M. Cotte, F. d'Acapito, L. Barni, C. Grazia, L. P. Buemi, J.-L. Andral, C. Miliani and A. Romani. 2022. "Deeper Insights into the Photoluminescence Properties and (Photo)Chemical Reactivity of Cadmium Red ($CdS_{1-x}Se_x$) Paints in Renowned Twentieth Century Paintings by State-of-the-Art Investigations at Multiple Length Scales". *European Physical Journal Plus* 137: 311.

Morales, K. M. and B. H. Berrie. 2015. "A Note on Characterization of the Cochineal Dyestuff on Wool Using Microspectrophotometry". *e-PreservationScience* 12: 8–14.

Munsell, A. H. 1905. *A Color Notation*. London: G. P. Putnam's Sons.

Nabais, P., M. J. Melo, J. A. Lopes, T. Vitorino, A. Neves and R. Castro. 2018. "Microspectrofluorimetry and Chemometrics for the Identification of Medieval Lake Pigments". *Heritage Science* 6: 13.

Nadal, M. E., E. A. Early and E. A. Thompson. 2006. *NIST Special Publication SP250-70: Specular Gloss*. Gaithersburg, MD: National Institute of Standards and Technology.

Nassau, K. 1987. "The Fifteen Causes of Color: The Physics and Chemistry of Color". *Color Research and Application* 12(1): 4–26.

Nassau, K. 1992. "Conserving Light Sensitive Minerals and Gems". In *Care and Conservation of Geological Material*, ed. F. Howie. Oxford: Butterworth-Heinemann.

Nassau, K. 2001. *The Physics and Chemistry of Color: The Fifteen Causes of Color*. 2nd edn. New York: Wiley.

Naumova, M. M., S. A. Pisareva and G. O. Nechiporenko. 1990. "Green Copper Pigments of Old Russian Frescoes". *Studies in Conservation* 35: 81–88.

Newton, I. 1704. *Opticks, or, a Treatise of the Reflexions, Refractions, Inflexions and Colours of Light*. London: Printed for Sam. Smith, and Benj.

Ormsby, B., R. Barker, M. Keefe, C. Tucker, F. Donate and P. Smithen. 2014. "The Removal of Graffiti Ink from Mark Rothko's *Untitled (Black on Maroon)* 1958: A Collaborative Approach". In *ICOM Committee for Conservation, 17th Triennial Meeting, Melbourne, 15–19 September 2014*, ed. J. Bridgland, art 1007. Paris: International Council of Museums.

Ostwald, W. 1916. *Die Farbenfibel*. Leipzig: Unesma.

Padfield, J. n.d. "National Gallery: Spectral Power Distribution (SPD) Curves". Available at: http://research.ng-london.org.uk/scientific/spd/ (accessed 16 August 2024).

Padfield, T. and S. B. Landi. 1966. "The Light-Fastness of the Natural Dyes". *Studies in Conservation* 11(4): 181–196.

Peggie, D. A., J. Kirby, J. Poulin, W. Genuit, J. Romanuka, D. F. Wills, A. De Simone and A. N. Hulme. 2018. "Historical Mystery Solved: A Multi-Analytical Approach to the Identification of a Key Marker for the Historical Use of Brazilwood (Caesalpinia Spp.) in Paintings and Textiles". *Analytical Methods* 10(6): 617–623.

Perkins, R., H. Owens, C. Carr, K. Hallett, I. Gibb and K. Frame. 2011. "Henry VIII's Tapestries Revealed". In *ICOM Committee for Conservation, 16th Triennial Meeting, Lisbon, 19–23 September 2011*, ed. J. Bridgland. Lisbon: Critério—Produção Grafica.

Pesme, C. and K. Dunne. 2022. "The Continuing Development of Display Recommendations at the National Galleries of Scotland: Exploring a Value-Based Decision Strategy for Light Risk Mitigation". *Studies in Conservation* 67(suppl. 1): 219–227.

Picollo, M., M. Bacci, A. Casini, F. Lotti, S. Porcinai, B. Radicati and L. Stefani. 2000. "Fiber Optics Reflectance Spectroscopy: A Nondestructive Technique for the Analysis of Works of Art". In *Optical Sensors and Microsystems: New Concepts, Materials, Technologies*, eds S. Martellucci, A. N. Chester and A. G. Mignani, 259–265. New York: Plenum.

Pretzel, B. 2008. "Now You See It, Now You Don't: Lighting Decisions for the Ardabil Carpet Based on the Probability of Visual Perception and Rates of Fading". In *ICOM Committee for Conservation, 15th Triennial Meeting, New Delhi, 22–26 September 2008*, ed. J. Bridgland, 759–765. New Delhi: Allied Publishers.

Price, K. E., C. Higgitt, T. Devièse, C. McEwan and B. Sillar. 2015. "Tools for Eternity: Pre-Columbian Workbaskets as Textile Production Toolkits and Grave Offerings". *British Museum Technical Research Bulletin* 9: 65–86.

Pridmore, R. W. 1999. "Bezold–Brucke Hue-Shift as Functions of Luminance Level, Luminance Ratio, Interstimulus Interval and Adapting White for Aperture and Object Colors". *Vision Research* 39(23): 3873–3891.

Radpour, R., G. A. Gates, I. Kakoulli and J. K. Delaney. 2022. "Identification and Mapping of Ancient Pigments in a Roman Egyptian Funerary Portrait by Application of Reflectance and Luminescence Imaging Spectroscopy". *Heritage Science* 10: 8.

Rae, C. and L. Sheldon. 2016. "Lost Purples: Detection, Composition and Reconstruction: Rediscovering the Appearance and Contemporary Resonance of Purple Paint in Tudor and Stuart Portraits". In *Colour Change in Paintings*, eds R. Clarricoates, H. Dowding and A. Gent, 103–109. London: Archetype.

Rawlins, F. I. G. 1936. "Studies in the Colorimetry of Paintings I". *Technical Studies in the Field of Fine Arts* 4: 179–186.

Richardson, C., D. Saunders and M. Spring. 2001. "Reconstructing Altered Colours by Image Processing: *The Conversion of the Magdalen* by Pedro Campaña". *Zeitschrift für Kunsttechnologie und Konservierung* 15(2): 338–346.

Riedler, R., C. Pesme, J. Druzik, M. Gleeson and E. Pearlstein. 2014. "A Review of Color-Producing Mechanisms in Feathers and Their Influence on Preventive Conservation Strategies". *Journal of the American Institute for Conservation* 53(1): 44–65.

Rowe, S. 2004. "The Effect of Insect Fumigation by Anoxia on Textiles Dyed with Prussian Blue". *Studies in Conservation*, 49(4): 259–270.

Roy, A. 1993. *Artists' Pigments: A Handbook of Their History and Characteristics*. Vol. 2. Washington, DC: National Gallery of Art.

Royce, K., C. Baars and H. Viles. 2023. "Defining Damage and Susceptibility, with Implications for Mineral Specimens and Objects: Introducing the Mineral Susceptibility Database". *Studies in Conservation* 68(3): 298–317.

Russell, J. and A. Donnithorne. 2014. "An Investigation of 'Faded' Metalpoint Drawings by Leonardo da Vinci in the Royal Collection". In *Leonardo da Vinci's Technical Practice: Paintings, Drawings and Influence*, ed. M. Menu, 267–282. Paris: Hermann Éditeurs.

Russell, W. J. and W. de W. Abney. 1888. *Report to the Science and Art Department of the Committee of Council on Education on the Action of Light on Water Colours*. London: Printed for H.M.S.O. by Eyre and Spottiswoode.

Saunders, D. 2020. *Museum Lighting: A Guide for Conservators and Curators*. Los Angeles, CA: Getty Conservation Institute.

Saunders, D. and J. Kirby. 1994. "Light-Induced Colour Changes in Red and Yellow Lake Pigments". *National Gallery Technical Bulletin* 15: 79–97.

Saunders, D. and J. Kirby. 2001. "A Comparison of Light-Accelerated Ageing Regimes in Some Galleries and Museums". *The Conservator* 25: 95–104.

Saunders, D. and J. Kirby. 2004. "The Effect of Relative Humidity on Artists' Pigments". *National Gallery Technical Bulletin* 25: 62–72.

Saunders, D., M. Spring and C. Higgitt. 2002. "Colour Change in Red Lead-Containing Paint Films". In *ICOM Committee for Conservation, 13th Triennial Meeting, Rio de Janeiro, 20–27 September 2002*, ed. R. Vontobel, 455–463. London: James and James.

Scott Taylor, J. 1935. *A Simple Explanation of the Ostwald Colour System*. London: Winsor & Newton.

Select Committee on the National Gallery. 1850. *Report from the Select Committee on the National Gallery, 25 July 1850*. Paragraph 669. London: National Gallery.

Spring, M. 2016. "Fading, Darkening, Browning, Blanching: A Review of Our Current Understanding of Colour Change and Its Consequences in Old Master Paintings". In *Colour Change in Paintings*, eds R. Clarricoates, H. Dowding and A. Gent, 1–14. London: Archetype.

Spring, M., M. Melchiorre di Crescenzo, C. Higgitt and R. Billinge. 2021. "Leonardo's *Virgin of the Rocks* in the National Gallery, London: New Discoveries from Macro X-Ray Fluorescence Scanning and Reflectance Imaging Spectroscopy". *National Gallery Technical Bulletin* 41: 68–117.

Standeven, H. A. L. 2008. "The History and Manufacture of Lithol Red, a Pigment Used by Mark Rothko in His Seagram and Harvard Murals of the 1950s and 1960s". *Tate Papers* 10. Available at: www.tate.org.uk/documents/376/tate_papers_10_harriet_a_l_standeven_the_history_and_manufacture_of_lithol_red_0.pdf (accessed 16 August 2024).

Staniforth, S. 1985. "Retouching and Colour Matching: The Restorer and Metamerism". *Studies in Conservation* 30(3): 101–111.

Stenger, J., N. Khandekar, R. Raskar, S. Cuellar, A. Mohan, R. Gschwind, M. S. Enriquez, C. Mancusi-Ungaro and C. Rosenberger. 2016. "Non-Invasive Colour Restoration of Mark Rothko's Harvard Murals Using Light from a Digital Projector". In *Colour Change in Paintings*, eds R. Clarricoates, H. Dowding and A. Gent, 73–76. London: Archetype.

Syme, P. 1821. *Werner's Nomenclature of Colours*. Edinburgh: William Blackwood.

Tamburini, D., C. Martin de Fonjaudran, G. Verri, G. Accorsi, A. Acocella, F. Zerbetto, A. Rava, S. Whittaker, D. Saunders and S. Cather. 2018. "New Insights into the Composition of Indian Yellow and Its Use in a Rajasthani Wall Painting". *Microchemical Journal* 137: 238–249.

Tanimoto, S. and G. Verri. 2009 "A Note on the Examination of Silverpoint Drawings by Near-Infrared Reflectography". *Studies in Conservation* 54: 106–116.

Thickett, D. 2018. "Frontiers of Preventive Conservation". *Studies in Conservation* 63 (suppl. 1): 262–267.

Thomson, G. 1986. *The Museum Environment*. 2nd edn. London: Butterworths.

Townsend, J. H., L. Carlyle, J.-H. Cho and M. Félix Campos. 2011. "The Yellowing/Bleaching Behaviour of Oil Paint: Further Investigations into Significant Colour Change as a Response to Dark Storage Followed by Light Exposure". In *ICOM Committee for Conservation, 16th Triennial Meeting, Lisbon, 19–23 September 2011*, ed. J. Bridgland. Lisbon: Critério—Produção Grafica.

Townsend, J. H. and R. Freemantle. 2023. "Three Works on Paper by Vincent Van Gogh: Technical Study, Display Considerations and a Conjectural Colour Reconstruction". *Journal of the Institute of Conservation* 46(1): 3–22.

van Loon, A., P. Noble and A. Burnstock. 2021. "Ageing and Deterioration of Traditional Oil and Tempera Paints". In *Conservation of Easel Paintings*. 2nd edn, eds J. Hill Stoner and R. Rushfield, 216–243. London: Routledge.

Verriest, G., A. Buyssens and R. Vanderdonck. 1963. "Étude quantitative de l'effet qu'exerce sur les résultats de quelque tests de la discrimination chromatique une diminution non sélective du niveau d'un éclairage C". *Revue d'optique théorique et instrumentale* 42: 105–119.

Waller, R. 1992. "Temperature and Humidity Sensitive Mineralogical and Petrological Specimens". In *Care and Conservation of Geological Material*, ed. F. Howie, 25–50. Oxford: Butterworth-Heinemann.

Watkinson, D. E., M. B. Rimmer and N. J. Emmerson. 2019. "The Influence of Relative Humidity and Intrinsic Chloride on Post-Excavation Corrosion Rates of Archaeological Wrought Iron". *Studies in Conservation* 64(8): 456–471.

West FitzHugh, E. 1997. *Artists' Pigments: A Handbook of Their History and Characteristics*. Vol. 3. Washington, DC: National Gallery of Art.

West FitzHugh, E. and L. A. Zycherman. 1992. "A Purple Barium Copper Silicate Pigment from Early China". *Studies in Conservation* 37(3): 145–154.

Whitmore, P. M. and G. R. Cass. 1989. "The Fading of Artists' Colorants by Exposure to Atmospheric Nitrogen Dioxide". *Studies in Conservation* 34(2): 85–97.

Whitmore, P. M., X. Pan and C. Bailie. 1999. "Predicting the Fading of Objects: Identification of Fugitive Colorants through Direct Nondestructive Lightfastness Measurements". *Journal of the American Institute for Conservation* 38(3): 395–409.

Williams, E. L., E. Grosjean and D. Grosjean. 1993. "Exposure of Artists' Colorants to Sulfur Dioxide". *Journal of the American Institute for Conservation* 32(3): 291–310.

Appendix A

Derivation of CIE Lab and CIE LCh data

$$L^{*} = 116f\left(\frac{Y}{Y_n}\right) - 16$$

$$a^{*} = 500\left[f\left(\frac{X}{X_n}\right) - f\left(\frac{Y}{Y_n}\right)\right]$$

$$b^{*} = 200\left[f\left(\frac{Y}{Y_n}\right) - f\left(\frac{Z}{Z_n}\right)\right]$$

where

$$f\left(\frac{X}{X_n}\right) - \sqrt[3]{\left(\frac{X}{X_n}\right)} \qquad \text{for}\left(\frac{X}{X_n}\right) > 0.008856$$

$$f\left(\frac{X}{X_n}\right) = 7.787\left(\frac{X}{X_n}\right) + 0.13793 \qquad \text{for}\left(\frac{X}{X_n}\right) \le 0.008856$$

$$f\left(\frac{Y}{Y_n}\right) = \sqrt[3]{\left(\frac{Y}{Y_n}\right)} \qquad \text{for}\left(\frac{Y}{Y_n}\right) > 0.008856$$

$$f\left(\frac{Y}{Y_n}\right) = 7.787\left(\frac{Y}{Y_n}\right) + 0.13793 \qquad \text{for}\left(\frac{Y}{Y_n}\right) \le 0.008856$$

$$f\left(\frac{Z}{Z_n}\right) = \sqrt[3]{\left(\frac{Z}{Z_n}\right)} \qquad \text{for}\left(\frac{Z}{Z_n}\right) > 0.008856$$

$$f\left(\frac{Z}{Z_n}\right) = 7.787\left(\frac{Z}{Z_n}\right) + 0.13793 \qquad \text{for}\left(\frac{Z}{Z_n}\right) \le 0.008856$$

and X_n, Y_n and Z_n are values of X,Y and Z for a perfect white, lit by whichever CIE standard illuminant is used in the calculation, for example, D65. See Table 3.3.

The chroma and hue in the CIE LCh system are given by:

$$C^* = \sqrt{a^{*2} + b^{*2}}$$

and

$$h^* = \arctan\left(\frac{b^*}{a^*}\right)$$

Appendix B

Derivation of the CIE 2000 colour difference (ΔE_{00})

$$\Delta E^*_{00} = \sqrt{\left(\frac{\Delta L'}{k_L S_L}\right)^2 + \left(\frac{\Delta C'}{k_C S_C}\right)^2 + \left(\frac{\Delta H'}{k_H S_H}\right)^2 + R_T\left(\frac{\Delta C'}{k_C S_C}\right)\left(\frac{\Delta H'}{k_H S_H}\right)}$$

where

$$\Delta L' = L'_1 - L'_2$$

$$\Delta C' = C'_1 - C'_2$$

$$\Delta H' = 2\sqrt{C'_1 C'_2}\ \sin\left(\frac{\Delta h'}{2}\right)$$

The subscripts 1 and 2 denote the values of L' or C' for the two samples under comparison, while the values of L', a' and b' are derived from L*, a* and b* as follows:

$$L' = L^*$$

$$a' = a^*(1+G)$$

$$b' = b^*$$

where

$$G = 0.5\left[1 - \sqrt{\frac{\bar{C}^{*7}_{ab}}{\bar{C}^{*7}_{ab} + 25^7}}\right]$$

with $\bar{C}^*_{ab}$ being the arithmetic mean of the C^*_{ab} values for the two samples, and

$$\Delta h' = h'_1 - h'_2$$

$$C' = \sqrt{a'^2 + b'^2}$$

$$h' = \tan^{-1}\left(\frac{b'}{a'}\right)$$

and the values of S_L, S_C, S_H and R_T are defined by

$$S_L = 1 + \frac{0.015\left(\bar{L}' - 50\right)^2}{\sqrt{20 + (\bar{L}' - 50)^2}}$$

$$S_C = 1 + 0.045\bar{C}'$$

$$S_H = 1 + 0.015\bar{C}'T$$

$$R_T = -\sin\left(2\Delta\theta\right)R_C$$

where

$$T = 1 - 0.17\cos\left(\bar{h}' - 30°\right) + 0.24\cos\left(2\bar{h}'\right) + 0.32\cos\left(3\bar{h}' + 6°\right) - 0.20\cos\left(4\bar{h}' - 63°\right)$$

$$\Delta\theta = 30\ e^{-\left[\frac{\bar{h}' - 275°}{25}\right]^2}$$

$$R_C = 2\sqrt{\left[\frac{\bar{C}'^7}{\bar{C}'^7 + 25^7}\right]}$$

Index

For Product Safety Concerns and Information please contact our EU representative GPSR@taylorandfrancis.com Taylor & Francis Verlag GmbH, Kaufingerstraße 24, 80331 München, Germany